A Volume in The Laboratory Animal Pocket Reference Series

The Laboratory
RAT

Second Edition

T0144206

The **Laboratory Animal Pocket Reference** Series

Series Editor
Mark A. Suckow, D.V.M.
Freimann Life Science Center
University of Notre Dame
South Bend, Indiana

Published Titles

The Laboratory Canine

The Laboratory Cat

The Laboratory Ferret

The Laboratory Guinea Pig, Second Edition

The Laboratory Hamster and Gerbil

The Laboratory Mouse, Second Edition

The Laboratory Nonhuman Primate

The Laboratory Rabbit, Second Edition

The Laboratory Rat, Second Edition

The Laboratory Small Ruminant

The Laboratory Swine, Second Edition

The Laboratory *Xenopus sp.*

The Laboratory Zebrafish

A Volume in The Laboratory Animal Pocket Reference Series

The Laboratory
RAT
Second Edition

Patrick Sharp • Jason Villano

CRC Press
Taylor & Francis Group
Boca Raton   London   New York

CRC Press is an imprint of the
Taylor & Francis Group, an **informa** business

Cover Credits (photos used with permission). Clockwise from top left: The Koken Rat®, Braintree Scientific, Inc.; high power magnification of *Aspiculuris tetraptera* ova demonstrating the typical egg morphology, Scott Trasti and Anna Acuna; Brown Norway Rat, Taconic Farms, Inc.; plastic, open top, primary enclosure, example of a plastic enclosure with a stainless steel wire-bar top, Allentown Caging, Inc.; Nude Rat, Taconic Farms, Inc.; *Ornithonyssus bacoti* (tropical rat mite), Lisa Heath.

CRC Press
Taylor & Francis Group
6000 Broken Sound Parkway NW, Suite 300
Boca Raton, FL 33487-2742

© 2013 by Taylor & Francis Group, LLC
CRC Press is an imprint of Taylor & Francis Group, an Informa business

No claim to original U.S. Government works

Printed in the United States of America on acid-free paper
Version Date: 20121023

International Standard Book Number: 978-1-4398-2986-8 (Paperback)

Library of Congress Cataloging-in-Publication Data

Sharp, Patrick.
    The laboratory rat / Patrick Sharp and Jason Villano. -- 2nd ed.
      p. ; cm. -- (The laboratory animal pocket reference series)
    Includes bibliographical references and index.
    ISBN 978-1-4398-2986-8 (softcover : alk. paper)
    I. Villano, Jason, 1979- II. Title. III. Series: Laboratory animal pocket reference series.
    [DNLM: 1. Animals, Laboratory--Handbooks. 2. Rats--Handbooks. QY 39]

636.088'5--dc23
                                2012024545

Visit the Taylor & Francis Web site at
http://www.taylorandfrancis.com

and the CRC Press Web site at
http://www.crcpress.com

To my father, James L. Sharp, and Jamie Weaver,
a childhood friend. "This one goes out to the one
I love... This one goes to the one I left behind" (REM).

Patrick Sharp

To Joyce, Josh, and Justin, and to my mentors.

Jason Villano

contents

preface, 2nd edition

The authors would like to thank Dr. Marie C. La Regina for her contribution to the 1st edition, which has served as a stable foundation for the current edition.

We have had the good fortune to have readers provide us feedback about what they wanted to see added in the 2nd edition. Thank you.

preface, 1st edition

The use of laboratory animals, including rats, continues to be an important part of biomedical research. In many instances, individuals performing such research are charged with broad responsibilities, including animal facility management, animal husbandry, regulatory compliance, and performance of technical procedures directly related to research projects. In this regard, this handbook was written to provide a quick reference source for investigators, technicians, and animal caretakers charged with the care and/or use of rats in a research setting. It should be particularly valuable to those at small institutions or facilities lacking a large, well-organized animal resource unit and to those individuals who need to conduct research programs using rats starting from scratch.

This handbook is organized into six chapters: Important Biological Features (Chapter 1), Husbandry (Chapter 2), Management (Chapter 3), Veterinary Care (Chapter 4), Experimental Methodology (Chapter 5), and Resources (Chapter 6). Basic information and common procedures are presented in detail. Other information regarding alternative techniques and details of procedures and methods beyond the scope of this handbook is referenced extensively, so that users are directed toward additional information without having to wade through a burdensome volume of detail here. In this sense, this handbook should be viewed as a basic reference source and not as an exhaustive review of the biology and use of the rat. The final chapter, "Resources," provides users with lists of possible sources of additional information, rats, feed, sanitation supplies, cages, and research and veterinary supplies. The lists are not exhaustive and do not imply endorsement of listed suppliers over suppliers not listed.

Rather, these lists are meant as a starting point for users to develop their own lists of preferred vendors of such items.

A final point to be considered is that all individuals performing procedures described in this handbook should be properly trained. The humane care and use of rats is improved by initial and continuing education of personnel and will facilitate the overall process of programs using rats in research, teaching, and testing.

acknowledgments

The authors wish to acknowledge the generous contributions of Dr. Emily Patterson-Kane, Dr. Timothy Cooper, Dr. James Griffith, Dr. Scott Trasti, Dr. Anna Acuna, Dr. Lisa Heath, Sherilyn Hall, Victoria Culbreth, Holly Kuepfert, Christina Gutting, George Vogler, Nikol Tschaepe, and the others credited in the book.

PES would like to give a shout out to the folks at the the Linuxcaffe (Toronto) as well as the West Coast Plaza (formerly Ginza Plaza, Singapore), St. Claire and Christie (Toronto), and the Belém Starbucks (Lisboa) for their permitting me to "zuccotti" while writing. Special thanks to Dave, Lena, Zhed, and Igor.

JSV would like to thank Dr. Ronald Wilson from the Department of Comparative Medicine, Penn State Hershey College of Medicine, and Dr. William Masters from the Animal Resources Center, University of Texas Medical Branch.

authors

Patrick Sharp received his Doctor of Veterinary Medicine (DVM) degree from Purdue University in 1991. He completed his postdoctoral fellowship at the Washington University School of Medicine in St. Louis and is a diplomate of the American College of Laboratory Animal Medicine (ACLAM). He is a Member of the Royal College of Veterinary Surgeons.

Dr. Sharp has worked in industry and has held domestic and international academic appointments and positions at the David Geffen School of Medicine at UCLA, University of Florida, the National University of Singapore, and the Fundação Champalimaud's Centre for the Unknown.

Dr. Sharp has supported comparative medicine training, including both ACLAM-approved training programs and international training opportunities for laboratory animal graduate veterinarians and veterinary students. He has strived to increase the quality of animal care through personnel education at all animal care levels and through the design and construction of better vivaria and research support facilities.

Jason Villano obtained his Doctor of Veterinary Medicine (DVM) degree in 2001 from the University of the Philippines. He worked as a clinical veterinarian at the Department of Experimental Surgery, Singapore General Hospital. During this time, he also completed his Master of Science degree in biomedical engineering at the Nanyang Technological University in Singapore. He completed his postdoctoral fellowship in laboratory animal medicine at Penn State Hershey College of Medicine. Dr. Villano has received his U.S. veterinary license and is an ACLAM diplomate. He is currently a clinical veterinarian at the University of Texas Medical Branch, Galveston, Texas.

important biological features

introduction

Taxonomy of the laboratory rat:
Kingdom: Animal
    Phylum: Chordata
        Class: Mammalia
            Order: Rodentia
                Suborder: Myomorpha
                    Family: Muridae
                        Genus: *Rattus*
                            Species: *norvegicus*

Rats are thought to have originated in the area of Asia currently occupied by Southern Russia and Northern China. *Rattus rattus* (black or ship rat, 2n = 38) was well established in Europe by AD 1100 and implicated in the bubonic plague that devastated most of Europe in the 1300s. *Rattus norvegicus* (brown rat, 2n = 42) was commonly found in Europe in the 1700s. This appearance followed thousands of years of absence. Fossilized rat remains dating to the Pliocene and Pleistocene epochs have been found in Europe. Until the writings of Giraldus Cambrensis (1147–1223), there was no distinction made between *R. rattus* and *Mus spp.* (mice). The late European arrival of *R. norvegicus* is offset by its ferocious nature, helping it to essentially eradicate the black rat from its former strongholds.

Fig. 1 Long-Evans, or Hooded, rat. (Image courtesy Taconic Farms, Inc.)

Fig. 2 Brown Norway rat. (Image courtesy Taconic Farms, Inc.)

Today, the black rat is restricted to areas near water, and the brown rat has conquered the planet owing to its climatic adaptability and ability to scavenge human refuse.

Today's laboratory rats are the domesticated descendants of *R. norvegicus* (**Figures 1** and **2**). Albino animals were held and used for rat shows, and frequent handling is thought to have tamed these animals. By the 1800s rats were used for breeding and neuroanatomical studies in the United States and Europe. Individual stocks and strains originated in the late 1800s and early 1900s. The laboratory rat has been and continues to be a biomedical research mainstay, with albino and pigmented animals available. There are recognized differences between wild and laboratory rodents. For example, laboratory rats have smaller adrenal and preputial glands, earlier sexual maturity, no reproductive cycle "season," better fecundity, and a shorter life span than their free-ranging wild counterparts.

nomenclature

Rats fall into two basic genetic groups: inbred or outbred. Inbred animals are generally classified into strains, and outbred animals into stocks (**Table 1**). An inbred strain is developed through at least 20 generations of brother–sister or parent–offspring matings to achieve an average 1% residual heterozygosity (alleles not fixed in the genome), whereas an outbred stock has less than 1% inbreeding per generation and has been maintained in a closed colony for at least four generations. Whether to use inbred or outbred rats in research is dictated by many factors, but usually it is by the study's purpose. For example, outbred stocks, such as Wistar and Sprague-Dawley (SD), are frequently used for general research when homozygosity is unimportant. Outbred stocks are also known for their hybrid vigor and good reproductive performance. In contrast, inbred strains such as the Fisher 344 (F344), Copenhagen, and Brown Norway (BN) rats, are used primarily for their well-defined genotypes and phenotypes related to the research subject. Different stocks and strains show variability in biological parameters, including hematology, clinical chemistry, behavior, infection resistance/susceptibility, and physiology (e.g., responses to anesthesia/analgesia and experimental procedures). Furthermore, because such variability occurs in stocks and strains from different suppliers, one should exercise great caution in changing suppliers mid-study. Examples include the genomic differences observed in Lewis and Wistar Furth substrains and the behavioral and neurochemical differences observed in SD rats. Canzian (1997) and Thomas et al. (2002) provide rat strain phylogenetic trees.

There are rules for rat nomenclature. Rat strain/stock name registration is with the Rat Genome Database (RGD). Outbred stocks are designated by the laboratory code followed by a colon and then

TABLE 1: SELECT RAT NOMENCLATURE EXAMPLES

Rat	Shorthand Designation	Stock/Strain	Coat Color
Lewis	LEW	Strain	Albino
Brown Norway	BN	Strain	Non-agouti brown
Fisher 344	F344	Strain	Albino
Spontaneously Hypertensive	SHR	Strain	Albino
(Charles River) Long-Evans	Crl:LE	Stock	White with black hood, occasionally brown hood
(Charles River) Wistar	Crl:WI	Stock	Albino
(Charles River) Sprague-Dawley	Crl:SD	Stock	Albino

stock designation in capital letters. For example, Crl:SD is Sprague-Dawley from Charles River Laboratories. Inbred strain nomenclature is more complicated and based on the rat's characteristic genotype. Here, we highlight the following:

- **Hybrid:** These rats are first-generation (F1) crosses between two different inbred strains. A FBNF1 rat is an F1 hybrid cross between a female F344 and a male Brown Norway rat.

- **Coisogenic:** These rats differ at only a single locus through mutation occurring in that strain. An example is the LEW.1AR1-*iddm* rat, a spontaneous diabetes mellitus animal model.

- **Recombinant inbred (RI):** These rats contain unique, approximately equal proportions of genetic contributions from two original progenitor inbred strains. RI strains are traditionally formed by crossing animals of two inbred strains, followed by 20 or more consecutive generations of brother x sister matings. An example of RI rats is HXB, derived from a cross of SHR/OlaIpcv x BN-*Lx*/Cub.

- Rats carrying induced mutations:
  - **Transgenic:** These rats carry a deliberately inserted foreign gene in their genome. An example is the HLA-B27 rat, a model of human inflammatory disorders expressing HLA-B27 and human β2-microglobulin (hβ2m).
  - **Knockout:** These rats have a single gene turned off through targeted mutation (gene trapping). Examples include the SD-*Foxn1*[tm1sage] and SD-*ApoE*[tm1sage].

- **Congenic:** These rats are produced by a minimum of 10 backcross generations to an inbred (background) strain, with selection for a particular marker from the donor strain. Congenic rat models include Type-2 diabetes mellitus, hypertension, and thymoma. Speed congenics, also known as marker-assisted breeding, selects progeny containing the highest percentage of recipient genetic background for further backcrossing. This strategy shortens the backcrossing to 5 generations, allowing the development of congenic strains in ~1.5 years.

- **Consomic:** Also called chromosome substitution strains, these are produced by repeated backcrossing of a whole chromosome onto an inbred strain. As with congenic strains, a minimum of 10 backcross generations is required. For example, in the SHR-Chr^{YBN} rat, the Y chromosome from the BN has been backcrossed onto the SHR.

Readers can consult guidelines of several organizations:

- Rat Genome Database, (RGD) Bioinformatics Research Center, Medical College of Wisconsin, 8701 Watertown Plank Rd., Milwaukee, WI 53226. Telephone: +1 (414) 456-8871; http://rgd.mcw.edu/.

- *Guidelines for Nomenclature of Mouse and Rat Strains.* Rat Genome and Nomenclature Committee and International Committee on Standardized Genetic Nomenclature for Mice; http://rgd.mcw.edu/nomen/rules-for-nomen.shtml.

- Institute for Laboratory Animal Research (ILAR), National Research Council, 500 Fifth St. NW, Washington, DC 20001. Telephone: +1 (202) 334-2590, Fax: +1 (202) 334-1687; http://dels.nas.edu/ilar. ILAR maintains the International Laboratory Code Registry. A laboratory code typically consists of 1 to 4 letters identifying a particular institute, laboratory, or investigator that produced, and may hold stocks of, a DNA marker, an animal strain, or a mutation.

behavior

Rats are nocturnal; thus reversing the light cycle in the animal room permits rats and investigators to share peak activity periods.

Although there are strain differences, rats are typically nonaggressive and will attempt to evade rather than confront. Males are usually more aggressive than females, and when striking, rats bite once. Rats feel most comfortable in small, dark, confined spaces, a behavior investigators may use as a reward. Like all rodents, rats are coprophagous, in that they ingest soft feces containing some parts of the orally ingested food and its metabolites. When designing experiments on drug metabolism and related studies, it is important to understand coprophagy and its potential impact on research and clinical chemistry results (e.g., increased blood urea nitrogen).

Rats are easily trainable and motivated to perform designated tasks by either positive or negative reinforcement. Since rats discriminate between familiar humans and strangers based on olfactory cues and choose to stay closer to familiar persons, frequent handling encourages the rats' non-aggressive nature as they adapt to new surroundings, including experimental situations. Water and/or food restriction, if scientifically justified, can also be used to train rats. Improper handling, nutritional deficiencies, and vocalizations of negative affected state (vs. positive, as in mating behavior) from conspecifics (e.g., 40-kHz vocalization when pups are separated from the dam) can result in undesirable rat behavior. Rats are inquisitive and may use either an active or passive strategy in coping with environmental demands. For example, they may escape or avoid aversive situations or may react with immobility, as in freezing behavior, an innate defensive reaction characterized by a lack of ambulatory movement with no or only slight head movement.

Perhaps the most studied aspect of the rat's behavior involves the role of housing. Interested readers are referred to the review paper by Balcombe (2006). Rats are generally not known to have stereotypies in the laboratory setting; however, impoverished cages can still lead to impaired brain development. Rats still benefit from an enriched environment, where they may hide, build nests, and make social contact. The possible importance of space to rats has rarely been studied, but the limited evidence suggests they value it. Rats could adapt well to individual housing, but male rats, unlike male mice, adapt quickly to social encounters and are unlikely to fight when housed together. However, an initial social encounter can still lead to fighting and stress, primarily to form dominant and submissive relationships. Several studies have shown that male and female rats suffer less stress from common husbandry and experimental procedures performed in the animal room when they are group housed than when singly housed. Other effects of isolation in the rat include reduced exploratory behavior, killing of young, mutual aggression, and altered biochemical parameters.

anatomical and physiological features

This section briefly summarizes the rat's anatomical and physiological characteristics. Special emphasis is placed upon those unique to the rat and relevant to its common research uses. **Table 2** summarizes the rat's basic biological parameters. In addition to the general organ weight information found below, Reed et al. (2011) provide organ weights (and body fat distribution) in various rat stocks and strains.

Like other rodents, rats have no sweat glands, cannot pant, and poorly regulate core body temperature. The development of thermoregulating mechanisms in neonatal animals is nonexistent until the end of the first week. Rats do not increase water intake at high ambient temperatures, with heat appearing to inhibit drinking. Relief is sought through behavioral means: increased salivation, burrowing, and shade. Adaptation to cold is better than to heat. Brown fat (multilocular adipose tissue, or "hibernating gland") is important in non-shivering thermogenesis during cold exposure. The rat's tail plays a role in thermoregulation, with the tail's vessels dilating to dissipate heat and constricting to conserve heat. A torpor-like decrease of core temperature in the morning occurs in juvenile Zucker (fa/fa) rats and

TABLE 2: BASIC BIOLOGICAL PARAMETERS

Parameter	Value
Lifespan (yr)	2.5–3.5
Adult male body weight (BW) (g)*	450–520
Adult female BW (g)*	250–300
Body temperature	35.9–37.5°C
	(96.6–99.5°F)
O_2 consumption (mL/m²/g BW)	0.84
Body surface area (cm²)	10.5 (BW in g)$^{2/3}$
Food intake (g/100 g BW/d)	5–6
Water intake (mL/100 g BW/d)	10–12
GI transit time (hr)	12–24
Urine volume (mL/100 g BW/d)	5.5
Total body water (mL)*/250 g BW	167
Intracellular fluid (mL)*/250 g BW	92.8
Extracellular fluid (mL)*/250 g BW	74.2

* Body weight (BW) will vary with stock or strain.

pigmented rat strains, such as the Long-Evans and Brown Norway. The thermoneutral zone is covered in more detail in Chapter 2.

Digestive System

Interested readers are referred to the reviews presented by DeSesso and Jacobson (2001) and Kararli (1995) for detailed descriptions and comparisons of the rat's digestive system with that of humans and other laboratory animals. Although the laboratory rat is widely used in digestion and nutrition research, current data generally indicates that no single animal model can mimic human gastrointestinal characteristics.

Oropharynx

- Dental formula: 2 (incisors 1/1; canines 0/0; premolars 0/0; molars 3/3) = 16.
- Like most rodents, the incisors' superficial enamel layer contains a yellow-orange iron-based pigment.
- Age at tooth eruption:
  - Incisors: 8–10 days
  - First molars: 18–19 days
  - Second molars: 21–25 days
- The incisors grow continuously; the molars do not.
- Lacks water taste receptors found in other animals.
- Lacks tonsils.
- The lack of canine and premolar teeth creates an open space termed the *diastema*.

Salivary glands

There are three pairs of salivary glands: parotid, submaxillary (submandibular), and sublingual.

Parotid

Each parotid gland consists of 3 or 4 lobes secreting a serous product. The glands extend ventrodorsally from behind the ear to the clavicle; the parotid duct opens opposite the molar teeth. The protein concentration (2%) of the saliva is unique, and the secretion volume is inversely influenced by dietary moisture content (the drier the diet, the higher the volume).

Submaxillary

Also called the submandibular glands, these are mixed glands with a serous and mucous product that dominates during grooming. They are found in the ventral region between the mandibles and thoracic inlet. There are two types of secretory granules found in the submaxillary glands, one in the acinar cells and the other in the granulated portion of the secretory ducts. Secretory duct granules of immature animals contain substructures not found in adult animals. Also, there is sexual dimorphism regarding the presence of mucous cells in the glands (100% in males but only 28.5% in sexually mature females).

Sublingual

The smallest of the salivary glands, they secrete a mucous product. The rounded glands are found at the rostral aspect of the submaxillary glands, and may be embedded in them.

Note: Brown fat is found in the ventral cervical region, and one should not confuse this structure with salivary glands or lymph nodes.

Esophagus

Like other rodents, the esophageal epithelium is covered with a keratin layer. The esophagus enters the lesser curvature of the stomach through a fold in the "limiting ridge" of the stomach. The fold prevents rats from vomiting.

Stomach

The rat's stomach has nonglandular and glandular portions separated by the "limiting ridge." The nonglandular forestomach's lining is similar to the esophagus and differs markedly from the glandular region, the corpus, which has secretory epithelium and prominent folds (rugae). In contrast to most mammals, the rat gastric mucosae secrete only pepsinogen C and cathepsins D and E, but not pepsinogen A.

Small intestine

Interested readers are directed to the discussions by Varga (1976) on the interaction of age, intestinal length, and transit time, and by Pacha (2000) on the developmental changes of intestinal transport. Generally, small intestinal lengths, transit times, and histological characteristics vary with the rat's age; Davis et al. (2011) found that the gastrointestinal transit time in 10-day-old SD rats was significantly less when the pups were separated from the dam. Average lengths for the duodenum, jejunum, and ileum in adult rats are 10 cm, 100 cm, and 3 cm, respectively. The numbers of intestinal villi decrease while that of the crypts increase with weight gain and age, especially at 21–35 days of age. There is also a decrease in villi size from the duodenum to ileum owing to the lower absorptive capacity in the last portion of the small intestine. Vacuoles with eosinophilic inclusion bodies are present within the enterocytes of the neonatal small intestine and are related to the milk antibody absorption. They have gradually disappeared by 18 to 21 days of age, presumably because of their association with the release of digestive enzymes and feeding changes. A unique rat small-intestinal feature is that the Brunner's glands are found only in the proximal duodenum. Carbohydrate absorption depends on physiological conditions, while amino acids and fats are absorbed primarily in the jejunum.

Large intestine

Cecum

The cecum is a thin-walled, comma-shaped pouch with a prominent lymphoid area found on the apex's lateral aspect. Although the rat cecum does not possess an inner septum as seen in other rodents, it has an inner constriction dividing the structure into apical and basilar sections. The lymphoid tissue is thought to be analogous to the vermiform appendix of human beings.

Colon

The colon has three divisions: ascending, transverse, and descending. The ascending portion has oblique mucosal ridges, whereas the mucosal folds of the transverse and descending regions have longitudinal mucosal folds.

Rectum

The rectum is that region of the gastrointestinal tract found in the pelvic canal.

Liver

- Liver weight: 10.0 g/250 g rat.
- Liver volume: 19.6 mL/250 g rat.
- Consists of four lobes: median, right lateral, left, and caudate.
- The rat has no gall bladder.

- The bile from each lobe leaves via ducts, which are unable to concentrate bile. These ducts then form the common bile duct, which enters the duodenum approximately 25 mm distal to the pyloric sphincter.

Pancreas

This is a lobulated, diffuse organ, extending from the duodenal loop to the gastrosplenic omentum. The pancreas has a darker color and firmer texture than the surrounding adipose tissue. The pancreas's diffuse nature results in a network of ducts, coalescing into 2–8 larger ducts emptying into the common bile duct.

Urinary System

- Kidney weight: 2.0 g/250 g rat.
- Kidney volume: 3.7 mL/250 g rat.
- The right kidney and adrenal gland are more cranial than their left counterparts.
- Rats, like other rodents, possess a unipapillate kidney, consisting of one papilla and one calyx.
- Long- and short-looped nephrons are present. The percentage of the latter and the relative medullary thickness of the rat kidney result in their ability to concentrate the urine twice that of humans.
- Rats are the only animals whose kidneys contain significant amounts of L-amino acid oxidase, an enzyme catalyzing the oxidation of 13 amino acids.
- The kidney also contains glutamine synthetase, an enzyme converting ammonium glutamate to glutamine. This enzyme is found in the kidneys of some animals but in all vertebrate brains.
- The female urethral orifice is at the base of the clitoris.

Integumentary System

- The rat epidermis is thinner than that of humans, but has a greater density of hair follicles. Thyroid hormone is important for promoting hair growth, with thyroidectomy causing hair growth retardation and epidermal thinning.
- The fur of albino rats tends to yellow with age.

Reproductive System

To distinguish between males and females, note that males have greater anogenital distance than females, and larger genital papillae (**Figure 3**).

Female

- Six pairs of mammary glands: 3 thoracic, 1 abdominal, 2 inguinal.
- Mammary tissue extends throughout much of the subcutis of the sides and necks.
- Duplex uterus consisting of two uterine horns, two cervices, and one vagina.
- The clitoris is the only genital structure connected to the urinary system.
- Clitoral glands are present and are analogous to the male preputial glands.

Male

- The inguinal canal remains open throughout the animal's life, enabling the male rat to retract the testes into its abdomen.
- No nipples.

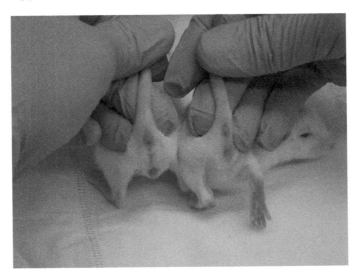

Fig. 3 Note the greater anogenital distance in the preweanling Sprague-Dawley male rat (left) compared to the female (right). It is best to make the comparison between litter mates. Male rats lack nipples.

- Has a *baculum* (an os penis, or penile bone), a single cartilaginous structure that supplies rigidity for copulation.
- Has a vascular penis with the body of the penis having 3 erectile components: 2 corpora cavernosa located dorsolateral to the urethra and a corpus spongiosus surrounding the urethra.
- Penile erections in young rats are known to occur around 30–40 days, and penile tissues show a very significant increase of NADPH (nicotinamide adenine dinucleotide phosphate diaphorase)-positive granules in the corpus cavernosum and the dorsal penile nerve during this period. NADPH is an enzyme involved in the synthesis of nitric oxide (NO), a mediator of smooth muscle relaxation in penile tissues. Thus, NO seems important for the development of penile erection in juvenile rats.
- Has 5 pairs of accessory sex glands: ampullary glands formed by the glands of the ductus deferens, seminal vesicles, prostate glands, bulbourethral (Cowper's) glands, and coagulating glands. The seminal vesicles and the coagulating glands secrete fluids that form the copulatory plug. The secretion of the Cowper's glands serves to clear the urethra of urine, to lubricate it and the vagina, and provide an energy source for sperm cells.
- The preputial glands, located towards the end of the prepuce, secrete smegma.

Respiratory System

- Normal respiratory function values are found in **Table 3**.
- Trachea consists of 18–24 rigid C-shaped cartilage structures.
- Lung weight: 1.5 g/250 g rat.
- Lung volume: 2.1 mL/250 g rat.
- Rats are obligate nose breathers.
- The left lung consists of one lobe. The right lung consists of four lobes: cranial, middle, accessory, and caudal. The middle lobe is also called the postcaval lobe because it is notched to accommodate the caudal vena cava, as it lies in contact with the diaphragm and apex of the heart. Lobectomy of the middle lobe can be performed safely and without detrimental effects on the rat.
- Bronchial constriction is under the control of the vagus nerve, not the adrenergic nerve supply.
- The lungs are immature at birth and consequently devoid of alveoli, alveolar ducts, and respiratory bronchioles. Gas exchange

TABLE 3: **RAT RESPIRATORY FUNCTION VALUES**

Parameter	Value
Tidal volume (mL)	0.6–2.0
Respiratory rate (breaths per minute)	70–115
Tracheal diameter (mm)	1.6–7.7
Minute ventilation (mL/minute)	75–130
Alveolar diameter (μm) [mean]	57–112 [70]
Total surface area (400 g animal, m²)	7.5
Thickness air-blood barrier (μm)	1.5
Alveolar length (μm)	288–624
Branches per alveolar duct	2–5
Atria diameter (μm)	15–262
Total lung capacity (mL)*	11.3 ± 1.4
Vital capacity (mL)*	8.4 ± 1.7
Functional residual capacity (mL)*	3.9 ± 0.8
Residual volume (mL)*	2.9 ± 1.0

Note: Surface area of the lung is related to O_2 consumption.
* 60–84-day-old anesthetized rats. The paper by Yokoyama (1983) presents values on older and younger rats.

occurs through the smooth walled channels and saccules until 4–7 days following birth, when remodeling occurs. Respiratory bronchioles are present 10 days following birth.

- There have been at least 10 morphologically distinct cell types identified in the intrapulmonary airways. Epithelial serous cell are thought to be unique to rats. The cells secrete a product that has a viscosity less than that of mucous cells and is thought to be responsible for the low-viscosity periciliary liquid layer found at all levels of the rat's respiratory tract.
- Regulation of the respiratory system occurs through tissue CO_2 exchange in the medullary respiratory center, with the carotid bodies playing a role. The carotid bodies, however, respond to low blood oxygen tension.
- The pulmonary artery is thinner and the pulmonary vein is thicker than most species. The thickness of the pulmonary vein is due to the cardiac striated fibers, which are continuous with those found in the heart and are observed mainly in the intrapulmonary branches. Unfortunately, this arrangement permits infectious agents to spread from the heart, through the pulmonary veins, into the lungs.
- Two types of pulmonary arteries are present: elastic and muscular arteries.

- As in humans, precapillary anastomoses occur in the lungs, and are limited to the hilar region. Pulmonary vasculature will vasoconstrict in response to acetylcholine (0.2–0.5 μg).
- Rats have a high neuronal density, high serotonin activity, and low histamine activity in the lungs.

Note: Papers by Stahl (1966) and Leith (1976) present methods for comparing respiratory variables between mammals.

Hemolymphatic System

- The thymus is located along the ventral aspect of the trachea, dorsal to the sternum, at the thoracic inlet. It involutes after puberty and may be unidentifiable as early as one year of age, particularly in males. The involution is under the control of β-adrenergic receptors, including those specific to glucocorticoids. However, recent studies reveal that thymic weight remains unaltered in old Wistar rats because the loss of functional tissue is compensated by an increase in the volume of interlobular connective and adipose tissue.
- Splenic hematopoiesis occurs primarily in the adult rat, especially in the presence of underlying infection. Thus, barrier-maintained rats should have minimal splenic hematopoiesis. Hemosiderin-laden macrophages are also normally seen in the spleen of female breeders.
- Hematopoiesis occurs throughout life in the long bones.

Circulatory System

- Normal cardiovascular values are found in **Table 4**.
- Heart weight: 1.0 g/250 g rat.
- Heart volume: 1.2 mL/250 g rat.
- The human and rat cardiovascular anatomy are similar, except for some notable idiosyncrasies. See the review by d'Uscio et al. (2000) for a detailed description of the cardiac physiology.
- Similar to that in fish, the cardiac blood supply originates from both the coronary and extracoronary arteries (internal mammary and subclavian arteries). The right coronary arteries supply the right and left atria, with the left cardiac arteries supplying a small portion of the left atria.

TABLE 4: CARDIOVASCULAR FUNCTION VALUES

Parameter	Value
Heart rate (beats per min)	250–450
pO_2 (mmHg)	82–94
pCO_2 (mmHg)	39.9
Arterial blood pH	7.41
H^+ (nM)	38.6 ± 0.6
Base excess	+1.8 ± 0.4
Arterial systolic pressure (mmHg) (mean)	88–184 (116)
Arterial diastolic pressure (mmHg) (mean)	58–145 (90)
Cardiac output (mL/min)	10–80
Blood volume (mL/100 g BW)*	5.6–7.1
Plasma volume (mL/100 g BW)*	3.08–3.67

* Values are approximate and can vary between sexes. For example, Probst et al. (2006) calculated blood volume in female SD rats to be 7.84 ± 0.70 mL/100 g compared with 6.86 ± 0.53 mL/100 g in males (14% gender difference). Plasma volume was also calculated to be 4.86 ± 0.54 mL/100 g in females compared with 4.12 ± 0.32 mL/100 g in males (18% gender difference).

- Another unique feature is the presence of two precavae in the rat vena cava, with the right precava emptying into the right atrium, and the left precava joining with the azygous vein before dumping into the right atrium.

- Mean flow rate may differ in contralateral arteries, as reported for the femoral arteries in Wistar rats. This physiologically significant difference is yet to be explained but might not be unusual given the mammalian anatomical asymmetry. Such cardiovascular findings have research implications, especially when using contralateral limbs as experimental controls.

- The oxygen metabolism and healing rate in rat tissues is faster than that of humans. The rates of vascular healing in rat carotid artery and aorta are qualitatively similar to those of larger species, although the vascular response in the rat, similar to that of mice, has only sparse thrombus formation and fibrin deposition, minimal inflammation, and a lack of remodeling. This potentially explains why research on antirestenotic compounds yields favorable results in rats and fails to produce similarly positive results in other animal models or in clinical trials.

- Heart rate (HR) and mean arterial blood pressure (MAP) increase in response to stressors, such as sepsis and witnessing or experiencing experimental procedures (e.g., restraint, injection). Such stress responses can be exacerbated during the metestrus–diestrus stage of the estrous cycle but can be minimized, not by increasing floor space per animal, but by group housing.

- The morphology of the rat's hematopoietic and circulating blood cells differs slightly from that of human blood cells. The basophil is the only human blood cell for which the rat does not have a readily recognizable counterpart.

- An important feature of rat platelets is that PAR3 and PAR4 are considered the primary thrombin receptors, as opposed to PAR1 and PAR4 in dogs and in humans.

Musculoskeletal System

- Vertebral formula: C7 T13 L6 S4 Cd27–30.

- The rat has five toes on each foot; the first digits of the fore-limbs lack the middle phalanx and are represented by flat-tened nails (*pollex*). The tibia and fibula are fused distally.

- Rat bone maturation is slower than in most mammals, with ossification being incomplete until after the first year of life. Recent studies indicate rat bone growth, compared with other species, continues much longer after sexual maturity and for a greater proportion of their lifespan. However, bone mineral density in 12-month-old rats is higher than that of 6- and 20-month-old rats. These considerations are important in using rats for bone or joint research.

- Rat bones, like those of mice and hamsters, do not have Haversian canals, and the skeletal epiphyses never close.

- The rat lacks true costal cartilages, as the dorsal segments of the ribs are usually calcified.

- There is considerable variation in bone structure, bone min-eral density, and fragility phenotypes between rat strains. This variability is strain- and anatomically site-specific, suggesting that no single gene regulates skeletal fragility at all sites.

- The thymus is believed to play a role in bone metabolism regulation. However, bone structure, function, and regenera-tive properties of athymic rats are no different from those of immunocompetent rats.

- Research studies have shown that calcitonin, parathyroid hormone, estrogen, and nitric oxide have important effects on bone functions.

- Androgen deficiency results in bone loss associated with significant osteopenia.
- Rat cardiac and skeletal muscles contain a significant amount of collagen.

Glands

- Lacrimal glands: Extraorbital lacrimal glands lay ventral and rostral to the ear, and intraorbital lacrimal glands lay in the caudal angle of the orbit.
- Steno's gland, found in the maxillary sinus, is homologous to the salt gland of marine birds. The gland regulates mucus viscosity and its watery, nonviscous secretion humidifies inspired air.
- Harderian glands, horseshoe-shaped structures in the caudomedial aspect of the orbit, are found behind the eye and produce a porphyrin secretion that usually goes unseen except in illness or stress. The porphyrin component causes crusts or tears. Porphyrin is often mistaken for blood, but porphyrin fluoresces red under ultraviolet light (Wood's lamp), while blood does not.
- Zymbal's gland is an auditory sebaceous gland found at the base of the ear.
- Adrenal glands are larger in females than males and in wild than domesticated rats.
- White adipose tissue secretes leptin, a hormone contributing to homeostasis by suppressing food intake and increasing energy expenditure in proportion to body fat levels. The Zucker rat is relatively insensitive to leptin and exhibits a phenotype characterized by excessive food intake.

Nervous System

The rat nervous system discussion here will be cursory in comparison to Paxinos's *The Rat Nervous System*, to which readers are referred for comprehensive information.

- Brain weight: 1.8 g/250 g.
- Brain volume: 1.2 mL/250 g.
- Lissencephaly is normal.

Central nervous system (CNS)

The brain (consisting of the cerebrum and cerebellum) and the spinal cord form the CNS. The CNS has three coverings, or meninges: dura mater, arachnoid, and pia mater. The dura mater is a tough fibrous material exterior to the other two meninges. The arachnoid is between the dura and pia mater, with the pia being significantly more delicate and thinner than the dura.

The brain and spinal cord are bathed in cerebrospinal fluid (CSF). Normative laboratory rat CSF values are shown in **Table 5**. The ventricular system is similar to other mammals, but the rat lacks a foramen of Magendie. The two lateral ventricles (first and second) feed the singular third ventricle via the interventricular foramen. Connection of the third and fourth ventricles is through the cerebral aqueduct. Although there is no median aperture communicating with the subarachnoid space, there are two lateral apertures. The spinal cord ends at the fourth lumbar vertebra.

Peripheral nervous system (PNS)

The rat's PNS consists of 34 pairs of spinal nerves arising from the intervertebral foramina of the cervical (8), thoracic (13), lumbar (6), sacral (4), and caudal (3) regions. Plexi form in the cervical, brachial, and lumbrosacral regions. The autonomic nervous system comprises the greater and lesser splanchnic nerves along with their plexi and ganglia.

TABLE 5: CEREBROSPINAL FLUID VALUES

Parameter	Value
Osmolarity (mOsm/kg)	302 ± 4
Na^+ (meq/L)	156 ± 2
K^+ (meq/L)	2.8 ± 0.1
Cl^- (meq/L)	126 ± 1
Glucose (mg/dL)	65
Formation rate (μL/min)	2.83 ± 0.18
TCO_2 (mmol/L)	27.4 ± 0.4
H^+ (nM)	44.3 ± 0.6
pH	7.35
Lactate (mmol/L)	2.8 ± 0.2
pCO_2 (mmHg)	48.5
CSF volume* (μL)	250 ± 16
CSF production rate (μL/min)*	1.88 ± 0.17
CSF pressure* (mmHg)	38 ± 4

* 30-day-old animals.

Special Senses

The review by Burn (2008) provides additional detail on this topic.

Vision

- The rat eye is exophthalmic but is otherwise generally typical of the mammalian eye.

- The rat has a retro-orbital plexus, in contrast to mice and hamsters, which have an orbital venous sinus. The plexus is formed by the external dorsal and ventral ophthalmic veins, and numerous anastomoses between these veins.

- Rats, especially albino rats, have relatively poor eyesight and narrower depth perception than humans. Thus, they rely heavily on facial vibrissae and smell for sensory input.

- Rats have two classes of retinal cones: one containing an ultraviolet-sensitive photopigment, and the other housing a pigment maximally sensitive in the middle wavelengths of the visible spectrum. Wavelength discrimination tests have indicated dichromatic color vision in the rat, contrary to the popular notion that rats are colorblind.

- Rat vision is more panoramic than binocular. They have a 40–75° binocular vision and a greater than 300° visual field. They also can see –10° and about +50° on the horizon, enabling them to see much above and below their heads without moving their eyes.

- Albino rats are sensitive to light (phototoxicity); light levels comfortable for humans can rapidly cause retinal atrophy and, in advanced cases, cataract formation. Retinal changes can occur in rats exposed to cyclic light intensity of 130 lux or higher at the cage level. However, some vision can remain after constant long-term light exposure, even when no intact photoreceptor cells can be observed. Even so, it is important to note that light-induced retinopathy can confound experimental procedures such as open field tests and the Morris water maze, where vision is often the most appropriate sense for guiding rats.

- Intraocular pressure (IOP) in the adult rat: 15–30 mmHg using an electric tonometer (Tono-Pen®); 13.0+/–1.2 mmHg using an induction/impact tonometer probe (rebound tonometry).

- Schirmer tear test: 10.2 ± 1.6 mm/min (Wistar rat); 5.8 ± 1.8 mm/min (Goto-Kakizaki [GK] rat, a spontaneous model of Type 2 diabetes).

Hearing

- The rat hearing range is approximately 0.25–80 kHz at 70 dB, with the greatest sensitivity between 8 and 50 kHz. Environmental enrichment vastly improves auditory neuron performance.

- There has also been a debate about whether rats can echolocate. Although there have been reports of blind rats being able to "echolocate," there is no evidence that rats can use sound to build up a detailed "picture" of their environment, as bats and cetaceans can.

- The range of good hearing is affected by the external ear (*pinna*), which can amplify or attenuate sound. The organ of Corti (*spiral organ*) of the inner ear contains auditory sensory cells (*hair cells*). There is direct correlation between the hair cell density, which decreases with age, and the auditory threshold, which is decreased by prolonged administration of kanamycin, an aminoglycoside.

- There are natural hearing differences across strains. For example, the F344 strain demonstrates approximately 20 dB better hearing at low frequencies (4 kHz), whereas the FBNF1 strain demonstrates approximately 20 dB better hearing at higher frequencies (32 kHz). Also, SD and F344 rats differ with respect to age-related hearing loss, although neither shows severe hearing loss as young adults. Thus, in sound exposure studies, not only is the choice of strain/stock important but also the standardization of the sound level.

Olfaction

- As mentioned above, the rat's sense of smell is well developed and is used for sensory input.

- Bowman's glands in the nasal mucosa produce a secretion bathing sensory receptors surfaces with an aqueous solution.

- The vomeronasal organ is a chemoreceptor organ with neurons detecting specific chemicals, such as pheromones, for communication signals.

- The olfactory epithelium is specialized for learned associations between volatile scents and their implications.

Touch

- Rats rely heavily on their vibrissae (whiskers), innervated by the trigeminal nerve. Each vibrissa is sunken in a follicle sealed within a blood sinus. When touched, the vibrissa bends and pushes the blood against the opposite side of the sinus, triggering a nerve impulse to the barrel cortex in the brain.
- As the main tactile organs, vibrissae (especially the mystacial vibrissae) are used to navigate, orient, and balance. Trimming vibrissae early in life leads to long-term sensorimotor function impairment despite training with regrown vibrissae.
- Rats can sometimes lack vibrissae for various reasons, such as barbering, which occurs much more commonly in mice. Nude rats completely lack whiskers, but most have short, kinked whiskers, giving a limited sensory range.
- Rats often bob their heads, which may help them gain motion cues about the distance of objects.

hematology and clinical chemistry

Since there exists extensive literature on the rat's clinical pathology, only a general overview is presented here. Readers are directed to *Clinical Biochemistry of Domestic Animals* (Kaneko et al., 2008), *Schalm's Veterinary Hematology* (Weiss and Wardrop, 2010), and *Clinical Chemistry of Laboratory Animals* (Loeb and Quimby 1999) for further information.

There are several factors to consider when interpreting rat clinical pathology results. In general, some analytes do not have the significance found in non-rodent species. For example, alanine aminotransferase (ALT) and sorbitol dehydrogenase (SDH) is liver specific in the rat, but aspartate aminotransferase (AST) is non-specific and of little diagnostic value. Recently, glutamate dehydrogenase (GLDH) and galactose single-point (GSP) method defined by the galactose blood level 60 minutes after intravenous galactose administration proved to be more sensitive liver biomarkers than AST or ALT. It is imperative for laboratories performing assays to establish in-house reference ranges and encourage submission of samples from control animals to ensure reliable results. Clinical chemistry tests are usually run on automated large-scale industrial clinical chemistry analyzers. Portable handheld clinical analyzers (e.g., i-STAT®) may also be used for some parameters, but it is advisable to first establish normal ranges. Variables impacting clinical pathology results include the following:

Stock/Strain: Certain analytes can be affected by a stock's/ strain's phenotype. For example, spontaneously epileptic rats (SER), because of their tonic convulsions, have increased blood urea nitrogen (BUN), ALT, and AST and decreased serum glucose.

Sex: Mature male rats have higher packed cell volume (PCV) and total leukocyte (lymphocytes and granulocytes) counts than females do. Male rats also have higher levels of cardiac troponin I (cTnI) and fecal corticosterone than females.

Age: Red blood cell (RBC) count, PCV, and hemoglobin levels are lower in juvenile rats than in adult rats. Relative and absolute neutrophil and lymphocyte numbers also vary with age. Presence of age-related diseases such as large granular lymphocyte (LGL) leukemia in F344 rats and spontaneous cardiomyopathy in male SD rats can also affect clinical chemistry and hematologic data.

Animal restraint and sample collection: Highly variable analytes such as lactate dehydrogenase (LDH) and alkaline phosphatase (ALP) can be affected by animal restraint methods and blood collection site and collection frequency. Repeated blood collection over 24 hours, corresponding to approximately 22% of the total blood volume, has been reported to decrease RBC parameters up to 30%, while a withdrawal of approximately 10% of the total blood volume can decrease RBC parameters by up to 10%. Adequate recovery to baseline blood values is proportional to the initial blood collection volume and is necessary with frequent/repeated sampling. Orbital puncture lowers K^+ levels and plasma luteinizing hormone-releasing hormone (LHRH). Orbital puncture also increases prolactin, corticosterone, creatine kinase (CK), and AST levels, suggesting this sampling method causes significant stress and tissue damage compared to using the tail or sublingual vein. Testosterone concentrations are decreased after ketamine/xylazine anesthesia and CO_2 euthanasia. Animal restraint and sample collection can unduly stress the rat, causing inaccurate values, including hematological parameters. Thus, refinement of procedures such as anesthetizing rats in their home cages (which can lower CK values and serum corticosterone) and acclimation to handling procedures can enhance assay accuracy and repeatability.

Other experimental and husbandry procedures: Such procedures can affect certain analytes, with most effects directly related to the stress induced by these procedures. Short-term stress causes the release of epinephrine, which induces increased glucose levels and lymphocyte counts, whereas long-term stress causes the release of corticosterone, which induces decreased counts of lymphocytes and eosinophils. Serum corticosterone increases in response to fasting and pen and large-group housing of rats. However, rats housed in solid-bottom polycarbonate cages with bedding and rats in standard wire-bottom caging have similar clinical laboratory data. Although with limitations, fecal corticosterone has been used as a noninvasive technique to monitor rat stress hormones. More recently, an enzyme immunoassay to monitor adrenocortical activity by measuring fecal corticosterone metabolites was validated. More immunoreactive fecal corticoid metabolites are detected in male rats, although female rats normally secrete more glucocorticoids into circulation.

Physiological and reproductive states: The estrus cycle phases affect analytes such as sex steroid hormones and gonadotropins. Meanwhile, plasma volume and arginine vasopressin increase during lactation and pregnancy. Other analyte values remain virtually unaffected until preparturition, when PCV, mean corpuscular hemoglobin concentration, ALP, albumin, total protein, glucose, total bilirubin, sodium, and chloride ion levels decrease and serum triglycerides increase. These analytes return to near-normal values as early as 1 day postparturition.

Sample handling and storage: Complete blood cell counts (CBC) should be analyzed the same day as collected. Storing properly collected EDTA-fixed blood samples at ambient temperature is possible and can still provide accurate hematological data for up to 48 hours after collection. Serum or plasma can be stored in a refrigerator or freezer, but some enzymes are unstable over long storage periods at any temperature. For example, refrigerating whole blood until centrifugation and serum separation reduces CK values. If storing samples is unavoidable, care must be taken to avoid sample evaporation by maintaining samples in appropriate containers. Inadequate mixing of stored samples may yield lower values for some analytes, such as albumin. Hemolysis invalidates many clinical chemistry tests by increasing the levels of most enzymes; hematology by decreasing PCV, platelets, and RBC values; and blood coagulation tests by increasing activated partial thromboplastin time (APTT), prothrombin time (PT), and clotting time because of decreased fibrinogen and coagulation factors.

In general, submit hematology samples in tubes containing the anticoagulant EDTA. Heparin is an unsuitable anticoagulant for determining CBC and interferes with Wright's stain for white blood cell differential counts. While most clinical chemistry assays are performed on serum, plasma may be acceptable. If EDTA is used to obtain plasma, K^+ and Ca^{++} chelation results in low levels of these analytes. When in doubt as to the type of sample to collect or the collection tube to use, contact the diagnostic laboratory first. **Table 6** provides rat hematology and clinical chemistry ranges.

Another commonly performed clinical laboratory test is **urinalysis** (**Table 7**). Urinalysis can be affected by various factors, including the physiology and anatomy of the rat urinary system. The amount of urinary protein is controlled by sodium depletion and local amino acid metabolism. Tubular α-globulin production is the source of the normally high rat proteinuria (0.4–1.0 mg/mL urine) compared to other

TABLE 6: RAT HEMATOLOGY AND CLINICAL CHEMISTRY VALUES

Parameter	Value
Packed cell volume (PCV)	35–57%
Red blood cell (RBC) count	$5–10 \times 10^6/\mu L$
White blood cell (WBC) count	$3–17 \times 10^3/\mu L$
Hemoglobin (Hb)	11–19 g/dL
Mean corpuscular volume (MCV)	46–65 fL
Mean corpuscular Hb concentration (MCHC)	31–40 g/dL
Mean corpuscular Hb (MCH)	18–23 pg
Reticulocytes	0–25%[a]
Platelets	$200–1500 \times 10^3/\mu L$
Neutrophils	13–26%[b]
Lymphocytes	65–83%[b]
Monocytes	0–4%[b]
Eosinophils	0–4%[b]
Basophils	0–1%[b]
Glucose	80–300 mg/dL
Alanine aminotransferase (ALT)	52–224 IU/L
Calcium	9.1–15.1 mg/dL
Phosphorus (inorganic)	4.7–16 mg/dL
Sodium	142–154 mEq/L
Potassium	3.6–9.2 mEq/L
Chloride	84–110 mEq/L
Blood urea nitrogen	11–23 mg/dL
Creatinine	0.4–1.4 mg/dL
Total protein	4.5–8.4 mg/dL
Albumin	2.9–5.9 g/dL
Total bilirubin	0.0–0.64 mg/dL

[a] Value is highly age related, with higher values being normal for weanling rats.
[b] Percent of total white blood cells.

TABLE 7: RAT URINALYSIS VALUES

Parameter	Value
Urine specific gravity	1.040–1.070
Urine pH	7.3–8.5
Osmolarity (mOsm/kg H_2O)	1659

TABLE 8: RAT COAGULATION VALUES

Parameter	Value
Activated partial thromboplastin time	15–19.3 sec
Activated clotting time	45.8–51 sec
Prothrombin time	15.7–25.4 sec
Thrombin time	23.1–32.7 sec
Fibrinogen	201.4–293.2 mg/dL

species. Proteinuria can also result from chronic progressive nephropathy (CPN) and should be distinguished from normally high proteinuria. The rat's osmotic ratio (urine/plasma) and urine specific gravity gives it a urine-concentrating ability twice that of humans. This difference in concentrating ability is due to the higher relative medullary thickness and the higher percentage of long-looped nephrons. The collection method and sample processing can affect urinalysis values. Cystocentesis provides the highest quality urine sample. Bacterial contamination is possible when employing other collection methods such as metabolism cages. Bacterial metabolism occurs when urine stands for prolonged periods at room temperature, shifting the urine's pH towards alkalinity. While freezing urine may be acceptable for some parameters, it is unacceptable for most microscopic analyses.

Coagulation tests include determination of PT, APTT, activated clotting time (ACT), platelet count, and fibrinogen level (**Table 8**). Typically the sample used for such tests is citrated plasma; however, the analysis of whole blood samples can also be performed. In one study, SD rats were found to have variably shorter ACT and lower sensitivity to heparin compared to humans.

reproduction

The following is a list of rat reproductive characteristics and reproductive values (**Table 9**):

- The **Bruce effect**, pheromones from a strange male preventing implantation, does not occur in rats.

TABLE 9: **RAT REPRODUCTIVE VALUES**

Parameter	Value	Parameter	Value
Puberty (d)	50 ± 10	Litter size (pups)	3–18
Vaginal opening (d)	28–60	Birth weight (g)	5–6
Estrous cycle (d)	4–5	Eyes open (d)	10–14
Postpartum estrus (hr)	3–18	Ears open (d)	12–14
Spermatogenic cycle (d)	12.9	Hair coat (d)	8–9
Maximum fertility (d)	100–300	Weaning (d)	21
Gestation (d)	21–23	Eating solid food (d)	11–13

- The **Whitten effect**, synchronization of estrus in females following the introduction of a male, although disputed is not thought to occur in rats.

- Induction of anestrous in some members of a group of females housed together is known as the **Lee Boot effect**. This does not occur as strongly in rats as it does in mice.

- Rats are continuous polyestrous.

- The estrous cycle is light sensitive; constant light results in persistent estrus and polycystic ovaries.

- Parturition lasts about 90 minutes.

- Postpartum estrus occurs, with about 50% fertility, and delayed implantation (4–7 days).

- Hemotrichorial discoid placentation.

- Breeding systems can be either monogamous (1 male, 1 female) or polygamous/harem (1 male, 2–6 females). The recommended minimum floor area of cages for a dam with a litter is 800 cm² (124 in²). Note that other breeding configurations may require more space and will depend on factors such as number of adults and litters, and size and age of litters.

- A copulatory (or vaginal) plug forms from semen coagulation following copulation. It fills the female's reproductive tract from the vulva to the cervix. It will remain for a few hours following copulation, and then will decrease in size and fall out. The role of the copulatory plug is not well understood but is suspected to be a reservoir for the gradual release of sperm or to prevent the outflow of sperm from the vagina. Pregnancy is rare without the plug.

- Outbred stocks are known for their good reproductive performance. In other rats, the origin and cause of altered reproductive performance is mostly unclear. For example, the poor reproductive performance of Dark Agouti (DA) rats, an inbred rat strain, is not associated with reproductive organ morphology, circulating testosterone level, or sperm production. However, DA rats have low testicular weight and variable epididymal sperm count, suggesting other factors in the testicular or epididymal environment alter male reproductive performance.

- There is a general belief that disrupting rodents postpartum will result in breeding failures, loss of newborn pups, parental neglect, or cannibalization and that some strains/stocks are perceived to have higher propensity to parental neglect following cage manipulation. However, changing cages postpartum has been shown to have no significant effect on the number of live pups weaned.

- Female rats exhibit regular cycles until "middle age," when animals transition to irregular cycles and finally to acyclicity by 12 to 18 months of age. Generally, by 10 to 12 months of age, animals lose the ability to conceive and are considered "reproductively senescent."

It is generally difficult to pinpoint the age of a rat's puberty onset but this could be as early as 25–34 days of age. Male rat puberty is marked by testicular decent into the scrotum and onset of spermatogenesis at about 45 days of age. However, optimal sperm production does not occur until about 75 days of age. In the females, vaginal opening, an apoptosis-mediated event occurring simultaneously with the first ovulation, is typically used as an external index of puberty onset. It occurs as a result of increasing estradiol secretion and can be stimulated by injection of estradiol into immature rats. Generally, changes in the uterus, including a sharp increase in uterus weight relative to the body weight, coincide with puberty.

Estrous Cycle

The estrous cycle is 4 to 5 days. Rats usually display regular cycles that are not disrupted easily, even with the routine stress in the animal facility. Adult female rats normally have cyclic uterine infiltrates of eosinophils. The estrous cycle phase is reliably determined by hormonal analysis and vaginal cytology either by direct examination ("wet smear") technique or with staining techniques (Papanicolaou and methylene blue) of vaginal smears. Wet smears use vaginal secretion collected by gently inserting the tip of a plastic pipette filled with 1 mL of normal saline (NaCl 0.9%) into the rat's vagina. This is

TABLE 10: VAGINAL CYTOLOGY CHARACTERISTICS

Stage	Duration (hr)	Cytology Findings and Notes
Proestrus	~12	Nucleated epithelial cells, leukocytes, occasional cornified cells
Estrus	~12	75% nucleated cells, 25% cornified cells; cornified cells will predominate as estrus progresses; high bacterial numbers
Metestrus	~21	Large numbers of leukocytes and cornified cells cellular debris; large, flat nucleated (pavement) cells; ovulation occurs during metestrus
Diestrus	~57	Consists mainly of leukocytes; low bacterial numbers

considered less time-consuming, more practical, and less expensive than using special stains. The characteristics of vaginal smears are shown in **Table 10**.

Vaginal impedance is an alternative method of determining the phase of the estrous cycle in rats, especially in distinguishing estrus from nonestrus rats. The inherent electrical impedance resistance of the vaginal wall changes predictably during the different estrous cycle phases, with the impedance of the epithelial cell layer of vaginal mucosa being significantly higher during the pro-estrus stage. There are several advantages to resistance measurements; however, lack of correlation between vaginal impedance and hormone levels may occur in some rat stocks/strains.

2

husbandry

Good animal husbandry is an important laboratory animal care component. Proper rat husbandry includes meeting the housing, enrichment, nutritional, and sanitation needs in a manner that minimizes stress to animals, caretakers, and scientific personnel. Other important aspects involve animal transportation and record keeping.

It is beyond the scope of this text to provide a deep discussion of animal facilities and their design and construction. Along with contemporary texts on the matter, there are several national meetings offering presentations and workshops on the topic. Meetings sponsored by groups such as TurnKey™ and Tradeline™ provide multiday presentations. The annual U.S. National AALAS meeting offers an unparalleled trade show accompanying AALAS' scientific program.

Note: Husbandry can be a significant experimental variable; therefore, direct veterinary oversight of the animal care program coupled with a highly-trained and valued husbandry staff are essential to minimize husbandry and other areas yielding research interference. Animal research facilities are consistently the most expensive infrastructure a research institution will invest and maintain. It is important for those operating the facilities to receive full institutional support, thereby enabling the institution to reap maximal benefits from its substantial investment and to have the facility and research support (e.g., husbandry) personnel function synergistically with the biomedical research program and personnel. Institutional administrators, especially the Institutional Official, must recognize this cost-value relationship and support the continually evolving animal care and use programmatic needs resulting from changing animal users, research programs, and regulatory requirements.

housing

Animal housing can be seen as five interrelated yet successively larger components or "environments": microenvironment, macroenvironment, megaenvironment, gigaenvironment, and teraenvironment. The microenvironment consists of an animal's cage or primary enclosure, while the animal's room constitutes the macroenvironment, with the building or facility being the megaenvironment. Contemporary biomedical research programs frequently consist of multiple facilities and campuses under an institutional umbrella supporting (i.e., financial, administrative, research, and local/national regulation aspects) those various facilities and campuses—the gigaenvironment—driving the animal research and research program needs. Because today's research environment is globally connected, animals, supplies/equipment, ideas, and people are more interconnected than ever, and therefore consists of a teraenvironment to support academic, government, and industrial research while meeting and understanding the demands that include an animal's infectious diseases, transportation requirements, and other specific legislation. In this text we focus upon the micro-, macro-, and megaenvironment.

Microenvironment

The rat's primary enclosure should address the following:

- Behavioral and psychological needs
- Visualization with minimal disturbance
- Social interaction and hierarchy development
- Adequate ventilation, food, and water
- Clean, dry, and safe housing

Other important primary enclosure considerations include cage size, design, and materials.

Size

Cage size is usually driven by regulatory mandate; however, concern in the laboratory animal community is growing because more science is needed to appropriately mandate cage sizes, and to some extent

TABLE 11: MINIMUM *GUIDE* LABORATORY RAT CAGE DIMENSIONS

Weight	Floor Area in.² (cm²)	Height in. (cm)
Rats in groups*		
<100 g	17 (109.65)	7 (17.8)
Up to 200 g	23 (148.35)	7 (17.8)
Up to 300 g	29 (187.05)	7 (17.8)
Up to 400 g	40 (258.00)	7 (17.8)
Up to 500 g	60 (387.00)	7 (17.8)
>500 g	≥70 (451.5)	7 (17.8)
Female rats + litter	124 (800)	7 (17.8)

* Consideration should be given to the growth characteristics of the stock or strain as well as the sex of the animals. Weight gain may be sufficiently rapid that it may be preferable to provide greater space in anticipation of the animals' future size. In addition, juvenile rodents are highly active and show increased play behavior. Larger animals may require more space to meet the performance standards.

cage materials. Clearly, the community is concerned about optimizing the animal environment, but wants to ensure that the space requirement, like other regulatory aspects, is appropriate and prudent. Cage space must be sufficient to allow the rat to turn around and make other normal postural movements. Although the scientific literature is replete with articles describing higher mouse caging densities, comparable rat information is currently lacking. The *Guide for the Care and Use of Laboratory Animals* (the *Guide*, **Table 11**), and regulatory mandates (e.g., the European Treaty Series *ETS 123* and the National Advisory Committee for Laboratory Animal Research (NACLAR) *Guidelines on the Care and Use of Animals for Scientific Purposes*) establish animal primary enclosure expectations. New to the *Guide*'s 8th edition are breeding space recommendations. The *Guide* states: "Other breeding configurations may require more space and will be dependent on considerations such as number of adults and litters, and size and age of litters. Other considerations may include culling of litters or separation of litters from the breeding group, as well as other methods of more intensive management of available space to allow for the safety and well-being of the breeding group. Sufficient space should be allocated for mothers and litter to allow the pups to develop to weaning without detrimental effects to the mother or the litter."

Design

The primary enclosure most frequently consists of a plastic "shoebox" cage with solid-bottom flooring (**Figures 4** and **5**) or (less frequently) a stainless steel wire-bottom (**Figure 6**). Although wire-bottom

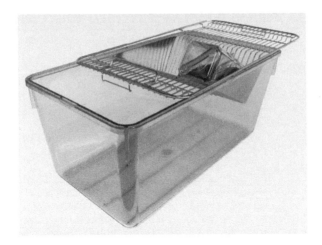

Fig. 4 Plastic, open top, primary enclosure. Example of a plastic enclosure with a stainless steel wire-bar top. The stainless steel top supplies an area for the water bottle and an adjacent space for food. (Image courtesy Allentown Caging, Inc.)

Fig. 5 Static microisolator, primary enclosure. The addition of the microisolator top reduces contamination from adjacent cages and the macroenvironment. (Image courtesy Allentown Caging, Inc.)

has advantages, rats prefer solid-bottom enclosures with bedding, and using wire-bottom enclosures is associated with foot lesions (e.g., ulcer/crust, callous, swelling); Peace et al. (2001) retrospectively determined that only those rats housed on wire-bottom floors more than one year were susceptible to developing foot lesions. Giral et al. (2011) found that rats housed even overnight on wire-bottom cages experienced an elevated heart rate, decreased locomotor activity

(a)

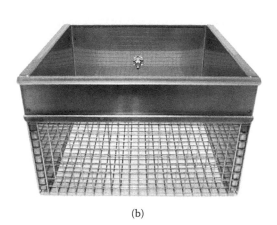

(b)

Fig. 6 Stainless steel wire-bottom cage rack assembly equipped with automatic watering (a), dimensions 1905 × 635 × 1880 mm. Individual stainless steel wire-bottom cage with automatic watering lixit (b), front view, dimensions 261 × 274 × 178 mm. (Images courtesy Allentown Caging, Inc.)

and BW gain, and a body temperature more subject to scotophase fluctuations.

The plastic, solid-bottom cage with the wire-bar top typically constitutes the basic unit; however, a further modification is the filter or microisolator top (**Figure 5**) to reduce the spread of various pathogens. This advantage far outweighs the additional handling time and elevations in temperature, humidity, gaseous, allergen, and particulate levels. The temperature and humidity increase can be minimized with good husbandry and cleaning practices, proper bedding selection, and provision of individually ventilated cages (IVC). Some research paradigms require elevating the animal(s) above the solid-bottom cage floor to minimize contact with excreta; stainless steel wire inserts for solid-bottom cages are available. When collection of urine and feces is necessary, metabolic cages may be employed, which also have a wire-bottom floor. Although the animals may be maintained in metabolic cages for only a short period, researchers should be aware that an IACUC may require scientific rationale when using any wire-bottom floor. Finally, researchers using research models requiring head caps may consider using automatic water and a hopper feeder (with a flat wire-bar top) or housing animals in transparent plastic 20L buckets to minimize the likelihood of the animals catching the head caps on the wire-bar tops used to hold food and water bottles; the buckets can be covered with the same REEMAY® filter material used on microisolator cage tops. Cage manufacturers can meet a host of unique research-related demands.

The best IVC designs bring rats into contact with high-efficiency particulate arrestance (HEPA) filtered air and offer intracage humidity, ammonia, and temperature reductions. Elevated ammonia levels may encourage disease development, (e.g., mycoplasmosis); therefore, cage designs and the associated husbandry practices must strive to minimize cage ammonia levels below 25 ppm (Broderson 1976). To preserve the integrity of the IVC's contents, the integrated cage (cage bottom and filter top) is transported to a HEPA filtered cage change station or BSC for any procedure requiring filter top removal; failure to do so negatively impacts animal and human health. Clearly some research practices (e.g., behavioral testing, bioimaging) cannot be contained within such a controlled environment, and these rats are at high risk for contracting the pathogens highlighted in Chapter 4.

Cage materials

Primary enclosures should be constructed from materials that are easy to clean and disinfect, durable, and corrosion-resistant. In addition, it is important for caging materials to be smooth, have impervious surfaces, and be easily kept in good repair. Stainless steel and durable plastics (most contemporary caging material being the latter) meet these criteria. Since the first edition of this text, concern over using plastic cages and their hormonal impact has been highlighted. Plastic primary enclosures may be constructed of polysulfone, polycarbonate, polystyrene, polyphenylsulfone, polyetherimide, polyethylene, or polypropylene construction. Polycarbonate permits easy animal visualization as it is the clearest, but polyphenylsulfone processing practices can yield a clarity rivaling that of polycarbonate. Plastic cages are subject to cracking, crazing, and fogging; when discovered, these cages require replacement.

Note: The estrogenic compound bisphenol A (BPA) is a widely used monomer that is polymerized to manufacture polycarbonate plastics. These plastics are used to manufacture caging for a wide variety of species, including rats. Concern exists over using some plastic cage products (e.g., cages and water bottles), most importantly damaged plastic materials, as these plastics may expose rats to BPA, an estrogen disruptor. BPA leaches from some plastics (e.g., polyphthalate) without visible damage and at room temperature. Damaged polycarbonate cages are the most likely culprit, especially when exposed to alkaline conditions, particularly at pH 12. Damage is most likely to occur when cages are hand washed using improper washing chemicals, and autoclaved. BPA's effects are modulated by dietary phytoestrogens. Although high dietary phytoestrogens provide protection from low-dose BPA exposure, the high phytoestrogen diet causes meiotic aberrations.

Although BPA is classified as a "weak" estrogen, it exhibits highly potent estrogen properties, even at low concentrations. Mouse exposure studies demonstrate BPA's effects on the fetal ovary by interference with estrogen receptor β. Research results vary depending on the rat stock/strain selected; CD-SD rats are reportedly insensitive. Other BPA exposure consequences include fetal oocyte meiotic defects, pregnancy complications, male and female reproductive tract abnormalities, and mammary and prostate gland morphology alterations.

In addition to BPA exposure, damaged cages risk compromising animal health, enhance escape, and inhibit adequate visualization.

The plastic primary enclosure's stainless steel wire-bar top contains the rats and provides a feeding area and a water bottle resting surface. Although adult rats can easily gnaw at the food between the bars, younger or debilitated rats may require other accommodations, including either placing feed or moistened feed on the cage bottom. Cages can accommodate various feeder styles and enable water bottle replacement without opening the cage's filter top. Specialized feeder designs facilitate the unique needs of various research projects, including a diet optimization feeder (Cover and Barron 1998) delivering a specified amount of ground diet to enhance animal survival on long-term toxicology and carcinogenicity studies. Animals on a liquid diet will require a modified feeder design and gnawing environmental enrichment to wear down the rats' continuously growing incisors.

Huerkamp and Dowdy (2008) evaluated diet replenishment ("topping up") compared with replacement with fresh chow in mice. They determined that mouse breeding pairs and group-housed mice consumed sufficient chow to support replenishment.

Contact bedding material (**Figure 7**), bedding in physical contact with the animals, should be absorbent, dust-free, nontoxic, inexpensive, sterilizable, contaminant free, and easily disposable. Generally, at least 6 mm of bedding material is appropriate. The contact bedding material impacts microenvironmental contaminants such as ammonia, especially those bedding materials containing urea-specific enzymes, or bacterial generation when the humidity exceeds the moisture threshold of a given bedding material. Above this threshold, urease-positive bacteria thrive, resulting in ammonia production. Contact bedding materials include recycled paper, ground corncob, cellulose, and wood chips (pine, aspen, or hardwood chips). Studies of various bedding types indicate corncob bedding, unbleached

Fig. 7 Contact bedding. Left to right: cellulose (Tek-Fresh®), aspen wood chip, corncob, rice hull.

eucalyptus pulp, and virgin cellulose maintain the lowest intracage ammonia concentration. Corncob bedding, when unautoclaved, has been associated with rat fungal rhinitis and with interfering with male and female rat sexual behavior through tetrahydrofurandiols (THF-diols) and leukotoxindiols (LDX-diols). THF- and LDX-diols occur naturally in corncobs and dietary corn products, with each compound having a lowest observed activity level of 0.5–1 ppm and 0.2–0.5 ppm, respectively. Previous contamination of wood with chemical (pesticides) and biologic agents (aflatoxins) can present a potential risk. Some wood products contain natural compounds that may alter research; specifically, cedar and some other softwoods contain cedrene and cedrol, which are known hepatic microsomal enzyme inducers. Davey et al. (2003) reported that animals exposed to pine contact bedding required several weeks for microsomal enzymes to stabilize following removal to wire-bottom cages in which there was no noncontact bedding substrate beneath the animals housed there. Failure to identify similar compounds in hardwoods (e.g., aspen) is responsible for their popularity. Some regions of the world have access to generous amounts of rice hulls, which yield comparable ammonia levels to corncob (Hua et al. 2009); bedding analysis from a Southeast Asian source generated concerns over the arsenic and lead levels found in the rice hulls, which requires appropriate testing and further evaluation before use. This contamination potential underscores the importance of periodic testing of any bedding substrate.

Cages may be autoclaved with or without bedding material, and cages must appropriately cool before introducing animals. Studies by Ward et al. (2009) indicate that bedding material clearly affects post-autoclave cooling by acting to retain heat after integrated (vs. nested) autoclave sterilization. Although the study determined that 3 hours was sufficient to introduce animals following autoclaving cages with ⅛-inch- (3 mm-) diameter corncob bedding 6 mm deep, each institution should determine an acceptable postautoclaving interval, as cooling is dependent on a host of factors, including autoclave manufacturer, autoclave settings, and the specific bedding materials.

Macroenvironment

The macroenvironment is the room or secondary enclosure (**Figure 8**). Factors affecting the macroenvironment include temperature, humidity, ventilation, illumination, and noise. Temperature and humidity are given joint consideration owing to their synergistic effects on heat loss and comfort.

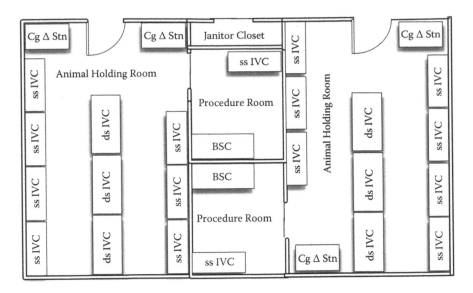

Fig. 8 Macroenvironment drawing highlighting the equipment orientation in the animal holding room; note adjacent procedure room for each holding room. ss IVC = Single-sided IVC; ds IVC = Double-sided IVC; Cg Δ Stn = Cage change station; BSC = Biosafety cabinet.

Temperature and humidity

The synergy of temperature and humidity impacts an animal's thermal balance. Recommendations for dry-bulb temperature and relative humidity for laboratory rats are 20–26°C (68–79°F) and 30–70%, respectively. Wild rats may require deviation from the laboratory rat recommendations in accord with their previous climate. Microenvironmental temperature and relative humidity may vary significantly from the macroenvironment, because of factors such as animal density, microisolator top use, bedding material, and cage-changing frequency. Increased animal density and microisolator use can elevate microenvironmental temperature and humidity. Certain bedding materials, such as rice hulls, corncob, and virgin cellulose, and an increased changing frequency result in reduced microenvironment humidity, and in turn lower ammonia production.

It is important to evaluate the macroenvironment temperature with respect to the animal's thermoneutral zone (TNZ), a temperature range where the metabolic rate and thermal stress are minimal. Various sources indicate the rat's TNZ, like the mouse's, is at the high end or above the *Guide*'s recommended temperature values. Reported TNZ values generally range from 26° to 29°C, and the actual preferred temperature may be determined by several factors, including physical environment, sex, age, and lactation. In comparing the TNZ between virgin and lactating rats, Roberts and Coward (1985) note that the lactating rats' TNZ was 8–20°C compared with 28–29°C for virgin rats. Concern exists that rats may be chronically cold stressed when exposed to ambient vivarium temperatures recommended in the *Guide* and other sources where the dry-bulb temperature is below the TNZ.

Refinetti and Horvath (1989) discuss a rat "thermopreferendum," "the range of ambient temperature preferred by the members of a species," of 19°C in a study using male Long-Evans animals. Other previously reported values found in the same article range from 22–32°C. While the thermopreferendum aligns closely with TNZ in mice and guinea pigs, this is apparently not the case in rats.

Ventilation

Microenvironment and macroenvironment interact most closely with respect to ventilation. Ventilation impacts the temperature, relative humidity, illumination (thermal load removal), and noise within the macroenvironment. The room's air supply must be sufficient for the number of primary enclosures and other research equipment. Ventilation not only provides an oxygen supply, but must also account for the thermal loads and moisture content within a room. Providing pressure differentials to maintain a negative pressure biohazard room relative to the hallway or to maintain a positive pressure barrier room relative to a hallway is yet another important role of ventilation. The general recommendation of 10–15 fresh-air changes per hour per room may result in over- or underventilation. One should therefore design air-handling requirements based on animals and equipment present in a room, being careful to accommodate for odor and allergen control, metabolic gases (including ammonia), and particulate debris. Generally, rats have the highest level of heat production of the commonly used laboratory animal species. The ventilation equipment itself will contribute to environmental noise and vibration through fans and air flowing through conduits; however, the effects can be minimized through proper facility design. Specialized equipment such as ventilated caging, Trexler flexible plastic-film isolators, and microisolators may further impact ventilation requirements.

Computational fluid dynamics (CFD) permits the study of air flow and movement and optimizes ventilation. CFD permits a user to evaluate various supply/exhaust options, animal loads, and equipment placement. While CFD has a cost, the end result is a more cost-effective ventilation arrangement minimizing entrainment, stagnation, and turbulence. Hughes and Reynolds (1997) and Curry et al. (1998) outlined the use of CFD in facility design projects to optimize ventilation designs in holding rooms and isolation cubicles.

IVC integration into renovation or new construction can help meet rapidly expanding colony needs. The article by Stakutis (2003) and the white paper by Phoenix Controls (2002) are good sources of basic information. It is important to carefully examine and balance matters such as the facility/research program needs, energy utilization, and life-cycle cost, and to understand how the IVCs will fit into the facility's emergency/disaster plan. The latter is an important consideration should the heating, ventilation, and air conditioning (HVAC) system fail while rack/tower blower motors continue operating. Clearly, this underscores the importance of an emergency/disaster plan, building-monitoring system, and prompt personnel notification.

Smart building-monitoring systems tailor a room's ventilation needs to its environmental quality. Sharp (2008, 2009, 2010) highlights systems monitoring laboratories, vivaria, and personnel areas simultaneously and responding with appropriate ventilation rates for the measured parameter(s) (e.g., particulates, ammonia) to bring the parameter(s) within acceptable limits. In addition to providing a performance-based indoor environment standard, the system provides significant energy savings over conventional systems, with a payback period as short as 1–2 years.

Illumination

Illumination must permit minimal interruption in the rats' behavior and physiology, yet enable lab workers to carry out routine animal care. The three most important aspects of illumination are spectral quality, photoperiod, and photointensity. The albino rat seems most sensitive to photointensity, and therefore most recommendations are based on phototoxic retinopathy. Acceptable light levels at the cage level range between 130 and 325 lux. Levels of 325 lux, measured 1 meter above the floor, are satisfactory for routine animal care. When animal care tasks require addition illumination, task lighting is available in cage change stations and BSCs. Photoperiod, or the light cycle, is commonly expressed as light:dark components; most research facilities operate on a 12:12 cycle, where the photoperiod is equally split with 12 hours of light and dark. Because photoperiod can significantly impact reproduction, some institutions use 13–14 hours of light for breeding colonies. Lighting intensity also can have an influence reproduction, with lower lux levels yielding better reproductive performance.

Timing devices must be checked regularly for proper function, as a longer (constant) photoperiod may result in the interruption of the normal estrous cycle. One should consider using automated data loggers (e.g., Onset Computer light on/off HOBO® data logger, **Figure 9**) positioned during the light cycle to minimize entering holding rooms during the scotophase (dark cycle). The lights being "on" during the scotophase (and vice versa) can have significant impact on research, and one must distinguish between a timing device failure and human error; the data logger's timing mechanism can assist in this determination. Facilities using overriding timers and switches should ensure that they cannot be tampered with or manipulated to permit more than the allotted light interval. Furthermore, as timers are frequently coupled with the use of red light, an appropriate film should be used (e.g., rose chocolate [Solar Graphics], Aegis® vivarium red film [Aegis Films]) to yield red light in a spectrum minimizing perception by the rat. Sodium lighting is a substitute for red lighting.

Fig. 9 HOBO® Light On/Off Data Logger. Dimensions: 2.4" × 1.9" × 0.8" (60 × 47 × 19 mm). Other data loggers are available to measure and store various environmental parameters (e.g., temperature, relative humidity) for later analysis. (Image courtesy Onset Computer Corporation.)

Natural, incandescent, fluorescent, and now light-emitting-diode (LED) lighting can provide illumination. Natural lighting in the form of hybrid lighting provides a constant light level without the variation in season cycles and intensity (e.g., cloud cover) or heat associated with windows. Lighting intensity and the associated phototoxicity discussed above are clear concerns with fluorescent lighting. Using LED lighting offers the following advantages over fluorescent lighting: no ultrasound emissions, no evidence of retinal damage while appropriately suppressing pineal melatonin, significant energy savings, and no mercury-associated hazards. The lack of retinal damage reported by Heeke et al. (1999) deserves further investigation and consideration for using LED lighting in view of its many advantages. Some facilities incorporate a dual light level with an automatic reset to the lower level after a given interval (e.g., 30 minutes). The dual level switches are associated with increased breeding performance, reduced phototoxicity, and reduced energy consumption.

Noise and vibration

Noise and vibration frequency (Hz) and energy (dB) can affect laboratory rats and the associated research. There are several papers highlighting circumstantial or retrospective evidence of nearby construction interfering with research results or animal breeding. Norton et al. (2011), Reynolds et al. (2010), and Rozema (2009, 2011) discuss the issues of quantifying noise and vibration in the vivarium. Construction materials (e.g., masonry walls) and practices (e.g., vibration dampening and isolation pads) can mitigate noise and vibration that may negatively affect research animals. Problems may appear localized to a room(s); however, they may be caused by equipment in the room (e.g., BSC) or the facility (e.g., HVAC system components). Fire alarms are another source of sound and illumination if equipped with strobe lights. To minimize a fire alarm's impact on research animals, the noise generated should be <1000 Hz and in the 70–100 dB range. Arrowmight Biosciences and Honeywell International Inc. manufacture vivarium-friendly alarms.

There are at least four noise and vibration sources in the animal facility: technical/mechanical devices, work in the room, work outside the room (e.g., construction in the building or neighboring buildings), and the animals themselves. The rat's hearing ranges from 500 Hz to 60–80 kHz, and plays a significant role in social behaviors. High-energy sounds and sounds occurring at irregular intervals and/or intensity may result in auditory and non-auditory changes in rats. Nonauditory changes include stress, metabolic alterations, and reduced fertility. Specific responses include increases in adrenal weights and total leukocyte counts and decreases in numbers of eosinophils (eosinopenia) and food intake. Individual behavioral differences correlate with corticosteroid elevations. Sound-insulating materials may conflict with the desirable construction qualities of laboratory animal facilities as they may not be impervious to moisture; therefore, placement of rats near loud animals (dogs, primates, etc.) and loud areas (cage wash, break rooms, etc.) should be minimized. Voipio (1997) has published the reaction of rats to specific sounds, including rat screams. Although there are no sound or vibration exposure limits or standards for laboratory animals, the U.S. Occupational Safety and Health Administration (OSHA), through the U.S. Department of Labor, regulates workplace safety and health, providing permissible human exposure standards.

Noise can be generated from various pieces of common equipment (e.g., fluorescent lighting, computers, squeaky castors) and nearby construction. Although high-frequency noise is quickly attenuated, lower-frequency noises and vibration associated with construction pass through the ground and buildings easily; the impact on ground level and subterranean vivaria may be greater than vivaria placed at higher levels. Fluorescent bulbs emit ultrasound, and the experimental impact on rats in the vivarium and laboratory is poorly understood. Equipment associated with such noise and vibration should be appropriately isolated, preferably close to the source. Equipment such as cage washers can have the pad isolated to minimize noise and vibration; however, construction noise (e.g., piling) is difficult to mitigate.

Co-housing rats and mice

Concern exists over co-housing rats and mice in the same room. Many vendors and some institutions co-house rats and mice of equivalent health status in the same room; vendors frequently use open-top caging, while institutions house animals in the range from open-top cages through IVCs. IVC use should minimize pheromone and pathogen transmission. The matter has been the subject of recent papers, with Meijer et al. (2009) and Pritchett-Corning (2009) finding no chronic physiological response or growth and reproduction effects, respectively. Arndt et al. (2010) found some significant physiological and behavioral differences; these animals were individually housed, and it is unclear what role this had in the research results. All three studies were conducted using open-top cages; IVCs might further diminish any pheromones or other odors.

Megaenvironment

The megaenvironment provides quality facilities where research may be performed. There are many sources of in depth information concerning facilities and facility planning. It is far beyond the scope of this text to go into details of facility design and construction. The following are recommended:

Texts/publications

- The Animal Welfare Act and Regulations: The definition of an animal "excludes … rats of the genus *Rattus* … bred for use in research…" Although laboratory rats are not a USDA-covered species, it would be the exception to have facilities designed solely for rats; therefore, general concepts should be followed.
- Good Laboratory Practice Standards (Sections 58.43, animal care facilities, and 58.90, animal care; U.S. FDA)
- *The Guide for the Care and Use of Laboratory Animals*
- *Public Health Service Policy on Humane Care and Use of Laboratory Animals*
- *The Handbook of Facilities Planning, Volume 2: Laboratory Animal Facilities*
- *The Laboratory Rat*, 2nd Edition (ACLAM series)
- *Planning and Designing Research Animal Facilities* (ACLAM series)
- *ALN Magazine*
- *NIH Design Policy and Guidelines*
- *Veterinary Medical Unit: VA Design Guide*
- Canadian Council on Animal Care Guidelines*: Laboratory Animal Facilities—Characteristics, Design, and Development*

Electronic resources

CompMed ListServe and CompMed Archives.

Meetings

The following meetings have a wealth of information on laboratory animal facilities and equipment. The AALAS meeting has perhaps the world's largest laboratory animal equipment exposition running in tandem with their excellent scientific program.

- National AALAS Meeting: www.nationalmeeting.aalas.org
- Tradeline Conference: www.tradelineinc.com
- TurnKey Conference: www.turnkeyconference.com

Five general areas of concern when building or renovating a facility are the location and design, construction and architectural finishes, facility monitoring, special housing and research needs, and security.

Location and design

Consideration should be given to many different issues when deciding where to put facilities. Among these are

- Special concerns: natural (e.g., earthquake, hurricane) and man-made incidents (e.g., break-ins)
- Geologic features of the proposed site(s)
- Climate: prevailing winds, building orientation, winter/summer severity, etc.
- Utilities and waste management
- State, local, and national regulations and codes
- Potential for future expansion, remodeling, and associated equipment movement
- Adjacent property utilization
- Accessibility and security
- Centralized vs. decentralized facilities
- Functional areas appropriate for the species, research paradigms, and research support program
- Species space requirements, key building support equipment (and redundancy), storage, etc.

Construction and architectural finishes

Room surfaces should permit ease of cleaning and be moisture and skid proof. Furthermore, the surfaces must withstand detergents, disinfectants, and temperatures used for cleaning and disinfection and permit rapid water removal. Wear and tear to the facilities can be minimized with proper placement of guardrails, corner guards, and kickplates. It is important to seal any cracks and penetrations to ensure adequate cleaning and not provide an area for vermin to hide.

It should be borne in mind how construction and finishes are appropriate to the species housed, and consideration should be given to maximal flexibility for the host of species that could eventually be housed. Furthermore, a dynamic research program may place additional utility, procedure space, or other future demands on the facility.

Facility monitoring

Monitoring and maintaining records on various facility parameters (temperature, humidity, photoperiod, etc.) help determine whether the facility is meeting the desired performance design standards. One may monitor electronically (computer records of data parameters), manually, or both. Monitoring is also a security issue to indicate after-hours access or alert personnel to dangerous parameters such as excessive temperature in a room. Facility access requires individuals to undergo training to understand and follow institutional practices and procedures for using the vivarium. Personnel entry requires some degree of personal protective equipment (PPE), with higher requirements in barrier and containment facilities; PPE must not be worn/used outside the vivarium. Facility entry requirements should be discussed with the facility manager or veterinarian, as these requirements may vary between institutional vivaria and are subject to change.

Special housing and research needs

The animals in an investigator's study may have special needs. These may include specialized housing (reverse light cycle, elevated room temperature, etc.) or research (cryopreservation, biocontainment, irradiator, imaging, etc.) needs. Housing needs may include a barrier system to maintain the animals' microbial status or minimize introducing undesirable agents. Research needs may involve using biohazard, physical, and/or chemical agents; preventing or minimizing unwanted animal and personnel exposure to these agents must be a priority.

Biohazard agents are assigned a biosafety level, which indicates the ease of contracting the agent, severity of the disease associated with the agent, and availability of treatment/vaccine. Identification of the animal biosafety level (ABSL) is through a numbering system (1, 2, 3, or 4). The biosafety-level rating depends on a given agent and the quantity of the agent under study. Studies involving agents in categories 1 and 2 can occur in barrier-type facilities, but those studies of category 3 and 4 agents require a higher level of containment. Further description of the biosafety levels and an agent's containment needs may be found in the U.S. Centers for Disease Control and Prevention's *Biosafety in Microbiological and Biomedical Laboratories* (the *BMBL*; 2009).

Using biohazard, physical, and/or chemical agents require appropriately designed facilities (e.g., room, equipment, etc.), personnel training, and signage. The signage serves to identify the hazards and as a written reminder to those previously trained, a training tool for new personnel, and warns those entering the facility during an emergency (e.g., fire, disaster) what precautions must be taken. Institutions should develop an Agent Summary Sheet highlighting:

- Agent/compound
- Specific PPE requirements
- Principal investigator, contact details
- Specialized equipment requirements
- EHS representative, contact details
- Bedding disposal
- Assigned room/location
- Waste handling

Housing methods

Rats are housed under at least one of these methods: conventional, barrier, and containment. The various housing methods may require personnel to follow a given procedure when entering the facility. These procedures may require some or all of the following:

- Disrobing
- Showering in and/or out of the facility
- Use of dedicated clothing
- Gowning
- Gloving
- Cap
- Mask/Respirator
- Booties

Conventional

Conventional housing generally consists of maintaining animals in open-top cages, manipulating animals/cages (including microisolator cages) outside a cage change station or BSC, and/or introducing unsterile items into the microenvironment (including feed, water, and bedding material). The benefits of microisolator tops and cage change stations/BSCs have been discussed previously. Under this definition, most vendors would be considered conventional; however, they employ varied practices that can shift the barrier from the microenvironment (seen in most research facilities) to the macroenvironment (or beyond).

Barrier

Generally, barrier facilities use microisolator cages and cage change stations/BSCs as part of an integrated disease prevention strategy frequently incorporating other methodologies, including autoclaving cage setups, water treatment, and autoclaving/irradiating diet.

Containment

A full discussion of containment is beyond the scope of this text. Containment provides animals and users protection from hazards including biologic, chemical, and toxic agents used in biomedical research. It is important to prevent agent cross contamination (to other animals, cages, and/or projects) and personnel/environmental exposure. It is important to employ barrier practices to prevent the animals from acquiring rat pathogens during a study, as this would confound research results. Containment frequently relies on engineering controls, facility design, and specialized equipment to prevent contamination of animals, personnel, and the environment; the equipment requirements depend on several factors and may include specialized caging, BSCs, and PPE. Work practices, SOPs, additional regulatory requirements, and staff training are significant containment investments and components.

Housing systems

Commonly used rat housing systems include shoebox cages (including open-top plastic and wire-bottom cages); static microisolators; and IVC, isolation, and disposable systems. This section emphasizes IVC, isolation, and disposable systems. Metabolic cages are a shoebox cage modification used to collect feces and urine (**Figure 10**); they are available as individual cages and racks.

IVC systems

IVC housing systems (**Figure 11**) have become extremely popular over the past several years and are a contemporary biomedical research mainstay; there are also containment IVCs available. They offer a high-density housing solution in various rack sizes to accommodate new construction and remodeling projects alike. There are

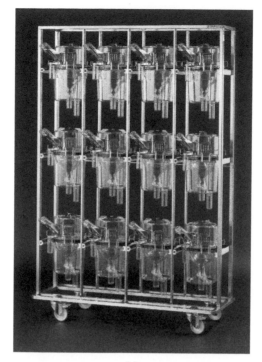

(a)

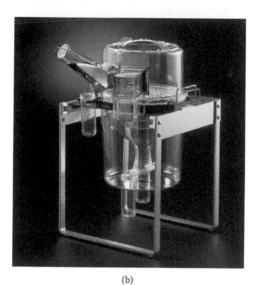

(b)

Fig. 10 A 12-unit metabolic cage rack (a). Dimensions: 1240 × 480 × 1900 mm. A single metabolic cage (b). Dimensions: 513–570 × 302 × 315 mm. (Image courtesy Tecniplast.)

(a)

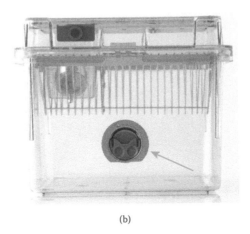

(b)

Fig. 11 (a) Single-sided 30-cage rat IVC rack equipped with 2 rack-mounted HEPA-filtered blowers and automatic watering. This rack's dimensions are 2127 × 699 × 1753 mm; other sizes, configurations, and capacities are available. The air plenums are located on the back of the rack. (b) IVC cage with a single, removable grommet (green arrow) providing both HEPA-filtered air supply (through a diffuser) and an automatic watering assembly. (Images courtesy Allentown Caging, Inc.)

many vendors, and attempts to classify one vendor as "superior" is fraught with complications, as studies often fail to ensure (among other things) proper IVC validation before experimental assessment. Concern also exists over the "high" air velocity that animals are exposed to, although "high velocity" is not quantified.

IVCs may use interstitial or room-based (e.g., tower, rack-/wall-mounted) blower configurations. Tower blowers (**Figure 12**) come with significant (opportunity) cost by occupying valuable floor space to provide ventilation compared to the alternatives. Rack-mounted blowers from reputable vendors are appropriately designed to prevent vibration

Fig. 12 Tower blower unit. Tower blower units are an alternative to wall, interstitial, and rack-mounted systems. The tower blower occupies valuable floor space that would otherwise house an additional vertical column of rat cages on an IVC rack. Dimensions: 330 × 658 × 1926–2086 mm. (Image courtesy Tecniplast.)

generation and transmission. Nonproprietary filters, Magnehelic® gauges, and the ability to perform in-house preventative maintenance significantly reduce an institution's recurring costs. Brown and Trent (2008) describe using an interstitial blower system to lower energy consumption by reducing the number of air changes per hour. Concern exists about CO_2 buildup in IVCs with solid plastic tops and those with a silicone seal between the cage top and bottom during power failure and animal transport in the cages (Nagamine et al. 2011); these cages are used extensively throughout Europe and in biocontainment IVCs. "Passive" IVC rack systems that attach to room exhaust are available; however, they do not offer HEPA filtration of room air immediately before and after introduction into the animals' cage.

Isolation systems

Providing isolation housing beyond the cage level may occur through various methods, including flexible film and semirigid isolators and cubicles (**Figures 13** to **15**). Flexible film and semirigid isolators

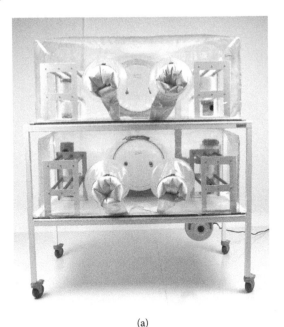

(a)

Fig. 13 (a) Front view, two-tiered flexible film isolator. Air is HEPA filtered; microisolator cage tops are thus unnecessary, with wire-bar tops preventing the animals from escaping. Cages rest on the gray racks, and manipulations are performed by individuals using the attached gloves. Dimensions: 1931 × 1677 × 909 mm.

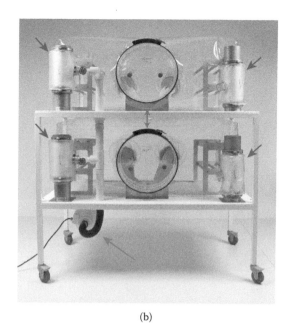

(b)

Fig. 13 (*Continued*) (b) Rear view, two-tiered flexible film isolator. The unit beneath the bottom shelf (green arrow) is responsible for moving HEPA-filtered air, with supply (red arrows) and exhaust (blue arrows) filters. The two large circular ports (orange double-headed arrow) are used to move materials in and out of the isolator. Dimensions: 1931 × 1677 × 909 mm. (Images courtesy Class Biologically Clean, Ltd.)

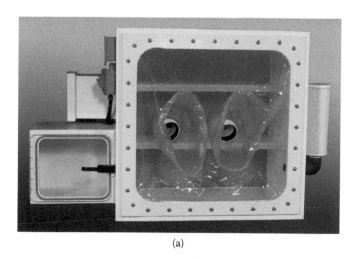

(a)

Fig. 14 (a) 3-ft (900 mm)-wide semirigid isolator.

(b)

Fig. 14 (*Continued*) (b) 6-ft (1.80 m)-wide semirigid isolator. (Images courtesy Park Bioservices, LLC.)

Fig. 15 "Illinois" or isolation cubicles with electronic vertical telescoping doors. (Image courtesy Britz and Company.)

provide isolation to animals frequently housed in open-top cages and were developed and used for axenic and gnotobiotic research. HEPA-filtered supply and exhaust air and careful work practices are employed to prevent the introduction or release of research-altering agents. Blower motor or electrical failure is a concern. Besides supporting axenic, gnotobiotic, and infectious-disease research, some institutions use these isolators to quarantine new arrivals or maintain important colonies. Animals are housed in cages with wire-bar tops. Staff training and the following of SOPs are essential to their use.

IVCs are used within the cubicle. Cubicles can be designed to mimic the ceiling drops of an interstitial blower system with rack blowers located (and accessible) in the cubicle's interstitial space. Units may include lighting; automatic watering; vertical telescoping doors with transparent, rat-appropriate red film or opaque windows; and internet/Bluetooth® building-monitoring system connectivity.

Disposable

Technically, all plastic cages are disposable; however, the cost of disposing of reusable IVC and static microisolator cages is prohibitive. Disposable cages are manufactured from polyethylene terephthalate. Primary cage and specialized disposable-only cage vendors are available with IVC compatibility. The cages and associated supplies are irradiated. These cages may be used by individuals experiencing cage wash or autoclave issues and select containment (e.g., biohazard, toxin, carcinogen) needs. The primary cage vendors have disposable cages and cage liners that fit into existing IVC racks. Concern exists regarding CO_2 buildup in disposable cages with solid plastic tops during power failure and animal transport in the disposable cages (Nagamine et al. 2011).

Security

Security systems are expected to alert appropriate individuals in the event of a break-in, mechanical failure, etc. They must be supplied with definite points of contact and must be incorporated in any disaster/emergency plan (e.g., "security tree," "telephone tree"). This plan is subject to periodic review and update. Security should be integrated into the building-monitoring system.

Besides locks and keys, one may wish to consider some type of card key or biometric access (e.g., fingerprint, retinal scan, palm biometry), as card keys and biometric devices will indicate who is accessing the facility and at what time, and thus can complement video monitoring. Card keys could still be passed between individuals or become lost. Mobile security guards and quality video monitoring (e.g., pan-tilt-zoom cameras, digital storage, periodic review) enhances security and serves as a deterrent.

An institution may integrate the security system into the institutional animal health program and limit access to a predetermined order of room or facility entry. This can serve to reduce the likelihood of disease transmission by affecting traffic and equipment flow. Although effective, this option requires extensive communication between the veterinary staff, animal users, and user committees.

Disaster and Emergency Planning

Disaster (and emergency) planning encompasses a wide range of activities and occurrences, including natural disasters, criminal activities, epidemic disease (e.g., H1N1 influenza), and other such matters impacting the facility's integrity or its ability to function appropriately. This is beyond the scope of this text, but there are several references which readers may find helpful: *ILAR Journal* 51(2) 2010, *Disaster Planning and Management*; *Veterinary Disaster Response* (Wingfield and Palmer 2009); and *National Earthquake Resilience: Research, Implementation, and Outreach* (2011). The AVMA has various disaster-related materials on their website. In addition, state and local officials may be a valuable information resource for disaster and emergency planning. While some disasters are immediate and unannounced (e.g., earthquakes), others have a "buildup" of several days (e.g., hurricanes). Effective planning must incorporate other institutional entities and their plans involving not only the incident but the aftermath as well. While planning is important, drills, "table-top exercises," and discussions with similarly affected individuals may identify weaknesses and alternative coping strategies. It is important to include disaster and emergency preparedness when considering the design and construction of new facilities and remodeling of older facilities.

In preparing for an epidemic disease (e.g., H1N1), animal care staff should receive a high priority for vaccinations or medical treatments, if available. Cross-training staff, developing animal "teams" (including both animal care and properly trained and select research staff), and other such measures ensure some consistency of animal care throughout the epidemic period.

Note: Public relations is key for a curious and skeptical public and every effort should be made to engage the institutional public relations team in preparation for a disaster, emergency, or other untoward occurrences. There are several state and regional resources to assist institutions (e.g., California Biomedical Research Association, CBRA; Mid-Continent Association for Agriculture, Biomedical Research and Education, MAABRE).

environmental enrichment

Environmental enrichment (EE) may be broadly classified into social and nonsocial. Social housing permits animals to act/react to non-specifics and express various rat-specific social behaviors. Rats, like other laboratory animals, prefer to interact with and manipulate environmental objects (**Figure 16**). Objects providing EE are usually an extension of the rat's normal behavior of gnawing and burrowing. It is important to realize that nests/nesting material may be appropriate for one species (e.g., mice) but requires caveats in rats where nest building is not innate. Rats provided nesting material from birth do learn to construct nests, making it a suitable environmental enrichment. FDA certified nylon cylinders were found to be a durable and cost-effective alternative to Nylabones.® Other enrichment includes:

- Wood blocks
- Plastic bones
- Shredded paper
- Social housing
- Stainless steel nuts
- Color marbles
- Plastic piping
- Flavored food pellets

Environmental enrichment has numerous benefits for animals; however, it can confound and complicate some types of research. Research depends on minimizing variables and taking a standardized approach to many environmental influences (e.g., nutrition, ventilation, illumination). Unfortunately, environmental enrichment's variables (age of initiation, frequency, duration,

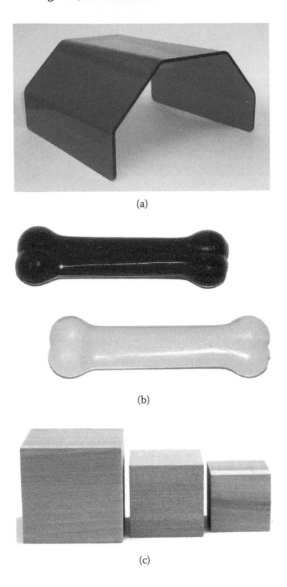

(a)

(b)

(c)

Fig. 16 Environmental enrichment examples. (a) Fat Rat Hut™ (159 × 86 × 152 mm); (b) Nylabones® (96–114 × 32 mm); (c) wood gnaw blocks (51, 38, and 32 mm³). (Images courtesy Bio-Serv.)

variation, complexity) are not standardized and are difficult to standardize; arguably this is both a blessing and a curse. EE's ability to confound research is perhaps most evident with behavior studies and is well discussed in the review by Simpson and Kelly (2011); similar confounding results have been reported in physiology and other types of research. Likewise, EE devices may complicate research owing to objects' variable levels of effective sanitation/sterilization, or by interfering in a study's objectives (e.g., nutrition, infectious disease, toxicology studies). Researchers must be aware of institutional EE policies and provide the necessary scientific justification to "opt out" of such policies if warranted; IACUCs may request pilot studies to further demonstrate specific research interference.

nutrition

Rat nutrient requirements will vary depending on this omnivore's life cycle, research goals, environment, microbiological status, and genetics. A nutritionally complete, uncontaminated, palatable, and properly stored diet should be available to meet these requirements. The National Research Council (NRC) of the United States establishes nutrient requirements for various animal species, including rats. The basis for the nutrient requirements is the lowest nutrient amount failing to show a deficiency, equilibrium between intake and excretion, and normal blood and urine metabolite maintenance levels.

Dietary Requirements

The rat's dietary requirements may be classified into five categories: energy, protein, minerals, vitamins, and other potentially beneficial dietary constituents. The specific nutrient requirements for maintenance, growth, and reproduction are published in the NRC's *Nutrient Requirements of Laboratory Animals* (1995 or newer). A specific concern to investigators is nutrient interactions affecting availability and husbandry issues such as animal housing. Housing that prevents coprophagy or uses galvanized (zinc) materials may significantly alter an animal's nutritional status. Coprophagy serves as a rat nutrient source, and preventing or restricting this activity may impact nutritional status. Using galvanized caging results in a lower zinc dietary requirement, as the animal may meet part of its requirement from gnawing on the caging. The NRC nutrient requirements discuss various rat-specific nutritional deficiency and toxicity signs.

Diets must be palatable; flavor and taste preferences have been evaluated in rats, primarily for rodenticide development. Of the four basic tastes, rats prefer sweet and salty, and generally reject sour

and bitter. Vitamin A deficiency reportedly alters rat taste preferences. Food preference is thought to develop from mother's milk
acclimation and by observing other rats. Not surprisingly, the rat
taste profile is homologous to that of humans. Neophobia is common
among rodents; novel-flavored diets may thus be initially rejected by
rats. Adding medication (e.g., fenbendazole) to the diets may affect
food intake and preference; fenbendazole-medicated feed (a pinworm
treatment) affected dietary preference, but not food intake, compared
to the standard, nonmedicated diet.

Energy

Rats are usually fed on an ad libitum basis and they consume a
diet to meet their energy requirement, which may vary depending on
temperature, age, physiology (e.g., lactation), and activity level. The
basal metabolic rate (BMR) in mature rats can be estimated using
the following formula:

$$H_{kcal} = 72\ BW^{0.75}$$

where
   H_{kcal} is the heat production in kcal per day
   BW is the body weight in kg
   72 is the average heat production (kcal) per $kg^{0.75}$

The equation to determine BMR in joules is

$$HJ = 301\ BW^{0.75}$$

The energy requirement may be met by feeding carbohydrates,
lipids, or protein; however, protein is an expensive way to meet the
requirement. In rats, this requirement is frequently met by dietary
carbohydrates and lipids. Although there is no specific carbohydrate
requirement, glucose, fructose, sucrose, starch, dextrins, and malt-
ose are the most common. It should be kept in mind that xylose is
toxic, and some carbohydrates may yield less than adequate performance. Lipids are a more dense energy source than carbohydrates,
a source of essential fatty acids (EFAs), assist fat-soluble vitamin
absorption, and increase diet palatability. The two EFAs of nutritional concern are the n-6 and n-3 fatty acids, known as omega-6
and omega-3 in popular literature. Linoleic acid will supply the n-6
fatty acids, and linolenic acid the n-3 fatty acids. Although the n-3
fatty acid requirement remains undefined, their inclusion in the diet
is advisable.

Caloric restriction

Numerous sources cite caloric restriction (30–40% restriction from ad libitum feeding) as increasing lifespan and life expectancy, decreasing degenerative disease incidence and severity, and delaying rat neoplasia onset. The concept is to reduce caloric intake without causing malnourishment. There are several theories to explain these results. In any event, employing ad libitum feeding in some studies (toxicology, aging) may warrant reconsideration.

In repeated survival and neoplasia studies, various dietary restriction levels (10%, 25%, and 40%) were evaluated against ad libitum feeding in male SD rats. The greatest marginal survival increase (24.1% at 110 weeks of age) occurred between ad libitum feeding and 10% dietary restriction; furthermore, the reduction in tumor occurrence in the 10% and 40% dietary restriction groups was virtually indistinguishable. In addition, no survival difference was observed between 10% and the commonly used 25% dietary restriction. Animals on the restricted diet received an NIH-31 diet formulated with 1.67 times more vitamins. This series of studies suggest that a lower dietary restriction (e.g., 10%) yields a result comparable to more drastic (e.g., 30–40%) caloric restrictions.

The beneficial effects of dietary/caloric restriction may result from hormonal (e.g., glucocorticoids) vs. cellular factors and may extend to other conditions such as cataract development, reduced inflammatory response, and reduced oxidative stress.

Proteins and amino acids

The existing 1995 NRC estimates recommend ~5% and 15% protein concentration for maintenance and growth/reproduction (females), respectively. Specific amino acid (AA) requirements are, on average, 23% higher than the pre-1995 NRC recommendations. Male rats fed a protein deficiency diet (3% and 10% protein) developed lower plasma testosterone levels (3% protein) and testicular weights (3% and 10% protein), and sperm were not observed in the epididymal duct lumens (3% protein).

Minerals

Minerals are divided into two categories—macrominerals and microminerals—based on their dietary requirement levels. There are six macrominerals of concern: calcium, phosphorus, chloride, magnesium, potassium, and sodium. The seven microminerals, or trace minerals, are copper, iodine, iron, manganese, molybdenum, selenium, and zinc. Mineral deficiencies can significantly alter research results. For example, diets deficient in magnesium are associated with behavioral abnormalities and altered ketamine and inhalant anesthetic requirements.

Vitamins

Vitamins may be classified as either fat-soluble or water-soluble. The fat-soluble vitamins are A, D, E, and K. The water-soluble vitamins are B6, B12, biotin, choline, folate, niacin, pantothenic acid, riboflavin, and thiamin.

Potentially beneficial dietary constituents

The 1995 NRC dietary recommendations do not have requirements for potentially beneficial dietary constituents; however, animals on natural ingredient diets appear to respond better to stressors than those animals on more purified diets. These potentially beneficial dietary constituents include fiber, chromium, lithium, nickel, silicon, sulfur, vanadium, ascorbic acid, and myoinositol.

Dietary types

There are three types of diets for laboratory animals, varying in their level of refinement: natural ingredient, purified, and chemically defined diets.

Natural ingredient diets

Natural ingredient diets are the most widely used diets owing to their low cost and high palatability. Because natural ingredient diets are relatively unrefined compared to the other diet types, nutrient concentrations may vary and contamination can occur. These two concerns may necessitate using purified or chemically defined diets in some studies. Open- and closed-formula diets are the two subclassifications of commercially available, natural ingredient diets. Open-formula diets specify the exact quantity of each dietary ingredient, whereas closed-formula diets do not. "Certified diets" undergo nutrient and contaminant analysis following preparation, an analysis is supplied with the diet, and the diets are frequently used for Good Laboratory Practices (GLP) studies.

Purified diets

The dietary ingredients in a purified diet are from a single nutrient or nutrient class. This level of refinement results in less nutrient variability and contamination than natural ingredient diets. These diets are less palatable and more expensive than natural ingredient diets.

Chemically defined diets

The dietary ingredients in chemically defined diets are individual AAs, specific sugars, chemically defined triglycerides, EFAs, inorganic salts, and pure vitamins. This level of refinement results in nutrient consistency and little contamination; however, significant concerns of chemically defined diets are poor palatability and high diet expense.

Dietary forms

Many dietary forms are available to the investigator; however, the most common is the pelleted diet. Other forms include extruded, meal, gel, crumbled, and liquid diets. Pelleted diets offer many advantages over the other forms with regard to handling, storage, use, and minimizing waste and dust. Meal and extruded diets are not commonly used because of waste. Extruded diets do offer the advantage of having a higher processing temperature, thus hindering microbial growth and viability. Dust generation and the need for specialized feeders limit this meal form's utility. Gel and liquid diets need refrigeration to reduce bacterial growth; however, owing to their form, they permit the addition of toxic or any other studied substance without dust formation. Müller (1992) describes a commercial elementary diet for long-term enteral nutrition in rats. Some dietary forms, especially gel and liquid diets, do not provide the "dental wear down" observed with pelleted diets, and unless gnawing is done by other means (e.g., wood blocks, nylon bones) malocclusion is likely to occur.

Dietary sterilization

Diet sterilization is a requirement for defined flora animals (e.g., specific pathogen-free, germ-free) and desirable for conventional animals. A diet may be sterilized by autoclaving, ionizing radiation (**Figure 17**), and filtration (e.g., chemically defined diets). Autoclaving results in greater pellet hardness and nutritional deterioration compared to ionizing radiation under ideal conditions (low moisture and packaging under vacuum or nitrogen). Since vitamins (especially B6) are especially vulnerable to damage during autoclaving, autoclavable diets contain 2–4 times the required levels to compensate for this.

Dietary contamination

Dietary contamination may negatively impact research projects, thus prevention is crucial. Contamination may occur well before diet manufacture. Several sources recommend acceptable levels of contamination (EPA 1979, FDA 1978, ICLAS 1987, Rao and Knapka

Fig. 17 Radura. This symbol indicates a product has undergone irradiation. While the shape is consistent, various colors are used.

1987). Contamination may be biological, chemical, or accidental, and fall into one of the following categories:

- Nitrates, nitrites, nitrosamines
- Pests
- Natural plant toxins
- Nutrient breakdown products
- Pesticides
- Heavy metals
- Bacteria, bacterial toxins
- Phytoestrogens
- Mycotoxins
- Formulation/manufacture errors
- Foreign materials, debris (e.g., glass, metal fragments, stones)

Phytoestrogens are naturally produced plant compounds capable of binding to estrogen receptors and exerting estrogenic/antiestrogenic effects. Soybean meal is a major rodent diet constituent and phytoestrogen source; alfalfa meal is also a phytoestrogen source. Researchers evaluating sex characteristics or differences should evaluate dietary phytoestrogen sources and consider alternatives (e.g., corn gluten meal) in consultation with a nutritionist.

Diet storage

Diet storage should be in a clean area to prevent vermin contamination and in a controlled environment to avoid environmental extremes. Store commercially available diets at <21°C and <50%

relative humidity, and purified or chemically defined diets at 4°C or as recommended by the manufacturer. Generally, use natural ingredient diets within 6 months of the milling date; more refined diets may have a significantly shorter shelf life (e.g., 30–45 days); low-temperature, low-humidity storage optimizes dietary shelf life. Rotate diet stock utilizing "first-in, first-out" management practices. Store opened diets in containers with tight-fitting lids, preferably marked with the milling date. Store diets, and other materials, on sanitizable pallets 6 inches (150 mm) from the wall; pallets on wheels greatly facilitate cleaning.

Water

Pelleted diets contain ~7–12% moisture, with water bottles, bags/pouches, or automatic watering systems providing the remaining ad libitum water requirement. Ambient temperature, relative humidity, and the diet's moisture content determine the rat's water requirement. Although each watering system has its own advantages and disadvantages, each is capable of supplying fresh water if adequate precautions are taken. Ultimately, it is important for the water reaching the animal to be drinkable and free of contamination. Ad libitum water availability is very important, as rats may consume ¼ to ⅓ of their body weight in water daily. For instance, at 22°C a rat's water consumption will outpace its food consumption by about 20%, and at 30°C water consumption will be about double food consumption on a per weight basis.

Although tap water may meet or exceed human consumption standards and regulations, water may need further treatment before being offered to a research animal. To control bacterial infections, such as *Pseudomonas aeruginosa* and *Pasteurella pneumotropica*, water acidification (pH 2.3–2.5) or chlorination (8–12 ppm) may be necessary. This specialized water treatment is an important consideration when using genetically modified (e.g., immunovague), immunocompromised, or immunodeficient animals in biomedical research, or for removing compounds from the water (e.g., the use of distilled water to reduce minerals in a nutrition study). Furthermore, it may be prudent to consider designing holding rooms without sinks to minimize the practice of inappropriately filling water bottles with regular tap water. One may acidify water by adding 0.8 mL of 25% hydrochloric acid to each liter of water and checking the resulting pH with a pH meter. Acidification at this level has no effect on body weight gain, hematology, biochemical analysis, or acid–base balance; however, it leaches minerals from rubber stoppers, especially zinc and chromium. Acidified deionized water will leach more minerals than deionized water, possibly impacting studies. It is also important to ensure that water treatment does not interfere with the research (e.g., enamel and dentine erosion when using acidified water for rats used in dental research).

It is important to periodically determine whether the water meets the research study's quality criteria, especially if there are any specific water treatments. Commercial laboratories are available that will perform water analysis.

One may present water via water bottles or automatic watering systems. Optimally, one should sanitize water bottles before refilling, and it is necessary to drain and flush automatic watering systems when sanitizing cage racks to minimize biofilm formation. Consider plastic over glass water bottles. Plastic bottles have the same BPA concerns as discussed above, but are lighter than glass; water bottles makes administering medication through the drinking water more efficient. When administering light-sensitive medication in the water, provide proper protection to the contents by using an appropriate bottle (translucent vs. opaque) or wrapping the bottle with aluminum foil. If automatic watering lixits are also available, remove the automatic watering lixits or place an approved lixit cap to ensure only medicated water is available to the animals. Horne and Saunders (2011) describe a carboy-based medicated water delivery system for automatic-water-equipped racks. It is important for the research and husbandry teams to clearly communicate (e.g., appropriate signs, stickers) on these special water projects to ensure that the specific cages are identified, that the duration of the water treatment is clearly stated, and that they have a laboratory contact telephone number. Signage indicating "Do Not Water" is inappropriate, as it appears the animals are to receive no water whatsoever.

Although automatic watering systems offer labor savings, daily inspections (just as those for water bottles) for inadequate or excessive flow are necessary; contemporary watering systems assess water flow and relay this information to building monitoring systems for appropriate action. Studies have found water bottles up to 13x more problematic than automatic watering (Rosa et al. 2010; Gonzalez 2011, 2008; Smith and Coleman 2010). Proper automatic water lixit handling, sterilization, and storage is critical; Latalladi et al. (2010) discusses using modified test tube racks. Automatic watering systems require periodic flushing with copious amounts of water or chemical agents followed by complete and thorough rinsing.

Lohmiller and Lipman (1998) discuss silicon crystal development and precipitation in tap water autoclaved in glass bottles. Although siliconized rubber stoppers and sipper tubes were not

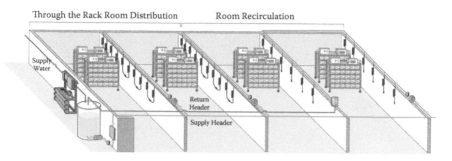

Fig. 18 Diagram demonstrating a recirculating automatic watering system. (Image courtesy Edstrom Industries, Inc.)

silicon contamination sources, autoclaving glass bottles with water siliconized rubber stopper pieces did result in silicon crystal formation by only a modest increase in silicon concentration.

Automatic watering systems pass municipally treated (institutionally untreated) incoming animal facility water through a series of filters and treatments; treatments include reverse osmosis and at least one of the following: acidification, chlorination, or UV irradiation. Treated water is distributed from a central supply reservoir via a recirculating (**Figure 18**) or nonrecirculating system (**Figure 19**).

Note: The authors strongly recommend incorporating automatic watering systems into the institutional disaster/emergency preparedness plan by placing them on emergency power and in a flood-resistant area of the building, with a suitably sized supply reservoir.

---

sanitation and sanitation monitoring

Proper sanitation includes the micro-, macro-, and megaenvironment, and restricts microbial growth, reduces animal disease risk, and reduces the potential for experimental variables. Sanitation impacts all animal facility areas, and the detergent and disinfectant choice must be a careful one while ensuring their use is according to manufacturer's recommendations.

Note: It is important NOT to mask odors. Odor masking is not a substitute for good sanitation and can result in physiological and metabolic alterations.

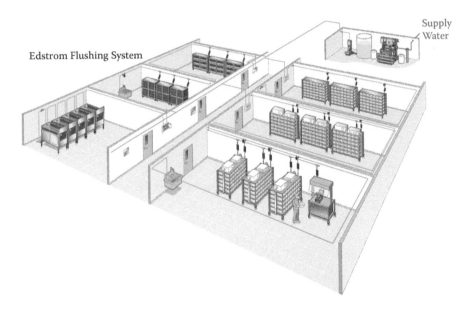

Fig. 19 Diagram demonstrating a nonrecirculating (flushing) automatic watering system. (Image courtesy Edstrom Industries, Inc.)

Sanitation of any item coming into contact with an animal is a logical, step-by-step process involving cleaning, washing/disinfection, rinsing, and sterilization. Although sterilization is highly recommended, not all facilities do so as a part of their sanitation program. Methods monitoring sanitation effectiveness include visual inspection, water temperature monitoring, and direct/indirect microbiologic assessment. Visual inspection ensures adequate physical cleaning before washing and disinfection. Physical cleaning reduces organic matter that may inactivate disinfecting agents. Monitoring water temperatures (58–82°C) ensures they are adequate for killing vegetative forms of pathogenic bacteria over a given duration (up to 30 minutes, depending on water temperature). One may monitor the temperature through either chemical indicators that change color or biologic indicators that contain organisms that are killed when an adequate temperature/duration combination is achieved (**Figure 20**). Microbiologic monitoring can be divided into direct (looking for growth of the whole, live organisms) or indirect (utilizing ATP, DNA, or other molecular surrogates) methods. Direct microbiologic monitoring ensures adequate sanitation through culturing a surface by using either a sterile swab or RODAC (Replicate Organism Direct Agar Contact) plates (**Figure 21**); however, results

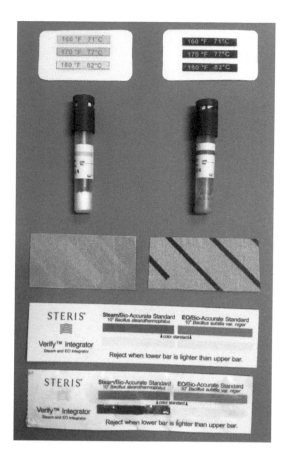

Fig. 20 Exposure indicators, from top to bottom, are cage wash self-adhesive temperature labels, autoclave biological indicators, autoclave tape, and a combined steam/ethylene oxide indicator. The left column and the top combined steam/ethylene oxide indicators are unexposed; the counterparts have been exposed (except for ethylene oxide) and demonstrate an appropriate "pass" reaction. The biological indicators require incubation following exposure. The remaining items are chemical indicators and can be read immediately. It is best practice to use the chemical indicators daily and incorporate the used labels into a logbook for the various pieces of equipment used for sanitation. The logbook should include a place for each exposed label, date, time, and initials of the person making the entry. The logbook must include a section for "Problems and Corrective Actions" when labels fail.

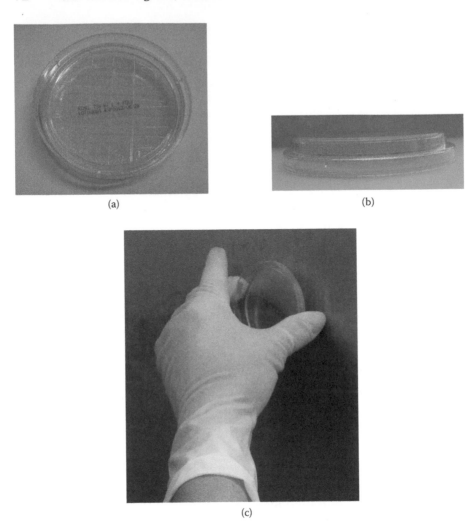

(a)

(b)

(c)

Fig. 21 (a) RODAC test plate showing a faint grid pattern, (b) the raised convex-surface media, and (c) the RODAC plate being applied to a vertical surface. The grid pattern assists in counting bacterial colonies that grow on the media after contact with a surface.

take time, arriving days after sampling. ATP-detecting molecular methods, such as luciferase testing, evaluate for ATP and provide immediate results (**Figure 22**). Nested PCR is another valuable, targeted molecular method used when identifying a select agent(s) in a disease outbreak situation; unfortunately results may be protracted, requiring several days. The various sanitation records should be kept, evaluated, and any sanitation problems/failures

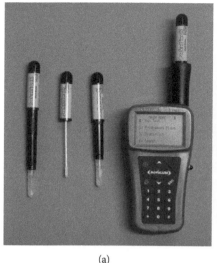 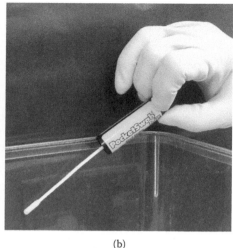

(a) (b)

Fig. 22 (a) Luciferase ATP reader and test swabs. (b) Sampling a surface.

investigated and remedied. These records, including those from the sentinel program, are an important component of an institution's Quality Assurance, as discussed below.

Sanitation chemicals must be compatible with the equipment on which it is used. Some chemicals corrode metal or other surfaces and negatively impact the integrity of the containers. An example of the latter is plastic chlorine dioxide spray bottles, which become brittle and break easily. Hankenson et al. (2011) reported on the loss of the integrity of nitrile gloves when exposed to chlorine dioxide for 60 minutes. To avoid corrosion, some surfaces may require a subsequent wipedown with alcohol. All sanitation chemicals must be appropriately labeled, including information on the contents, specific hazard (e.g., corrosive), formulation strength, and expiration date. Always follow manufacturer recommendations and use the necessary safety equipment.

Microenvironment

Cage changes should be frequent enough to keep rats clean and dry. This frequency is a function of physiological status, caging density, and experimental design. For instance, cages holding rats with diabetes mellitus or insipidus require more frequent cage changes than animals in a breeding colony, where frequent cage changes may alter pheromone concentrations. During bedding changes, it is wise to limit personnel exposure to particulates, aerosols, and allergens by using cage change stations/BSCs, and appropriate PPE.

- Sanitation of solid bottom cages and their accessories should occur at least weekly, and other primary enclosures (suspended stainless steel cages) every two weeks. Longer cage sanitation intervals may be acceptable, with supporting data, and are subject to IACUC approval.
- One should disinfect cages, following cleaning, with chemical agents (disinfectants) and hot water to destroy vegetative forms of pathogens.
- Detergents and disinfectants enhance the effectiveness of hot water, but necessitate thorough rinsing to avoid problems when animals and people come into contact with these agents.
- The concept of cumulative heat factor involves hot water temperatures—58–82°C (143–180°F)—and contact time. As water temperature decreases, the minimum contact time increases, and vice versa.
- Schondelmeyer et al. (2006) evaluated the sanitation frequency of rodent caging accessories (e.g., wire bar tops, filter material/support structure) using ATP levels and RODAC plates, and determined that ATP levels did not differ significantly between 14 and 90 days and bacterial colony counts had no significant difference between 14 and 120 days.

Although acceptable, hand washing cages and cage accessories is labor intensive and requires attention to detail. In addition, there is a greater risk of personnel exposure to a variety of hazards. A mechanical cage washer is preferable to manual cleaning and is required in some situations (e.g., ABSL 3 and 4). Mechanical cage washers may be broadly classified as cabinet, rack (**Figure 23**), or tunnel (**Figure 24**) washers. Tunnel washers offer a high degree of process automation by incorporating dump stations, cage handling (e.g., robotics, linear automation), and the placement of clean bedding in the freshly washed/dried cage.

Many facilities sterilize cages and cage accessories after cleaning and disinfection. Cages and cage accessories may be sterilized by

Fig. 23 High-energy efficiency, water-saving, high-throughput, pit-mounted cage washer. Exterior dimensions: 2337 × 2743 × 2477 mm. (Image courtesy Northwestern Systems Corporation.)

autoclave, dry heat (forced air convection), gas sterilization, or ionizing radiation; however, most institutions use large bulk autoclaves (**Figure 25**). Sterilization also requires a method of regular monitoring to ensure sterility using either contact or biologic indicators (see above).

Macro- and Megaenvironment

The rooms and the building require regular cleaning and disinfection schedules. Sweeping and mopping schedules are augmented by periodic cleaning and disinfection of walls, ceilings, and other surfaces and spaces within the macro- and megaenvironment. The equipment should be of sound, durable, noncorrosive construction, kept in good repair, and stored appropriately to avoid clutter and contamination and to permit drying. It is important to label any cleaning equipment with the room number where it is used to ensure it returns to that room following periodic cleaning/sanitation.

Fig. 24 High-energy efficiency, high-throughput, floor-mounted tunnel washer. Exterior dimensions: 1320–1930 × 2515 × 6096 mm. The tunnel washer is available in a variety of belt widths. (Image courtesy Northwestern Systems Corporation.)

Fig. 25 Large, bulk, pit-mounted sliding-door, pass-through autoclave. Autoclave chamber dimensions: 1830 × 2200 × 2190 mm. (Image courtesy Tuttnauer.)

Fumigation is a popular method of disinfecting equipment and holding rooms. There are several choices; vaporized hydrogen peroxide (VHP) and chlorine dioxide (CD) gas are the most common, as formaldehyde is a suspected carcinogen. Room fumigation may occur through specialized room modifications (e.g., fumigant gas injection ports, ventilation damper controls) or less specialized means to ensure appropriate gas containment to a given room/area (**Figure 26**). VHP has specific humidity and environmental requirements, a longer cycle time, is inactivated by certain metals, and is a vapor and not a true gas; high humidity limits the amount of VHP that can be vaporized and maintained in an appropriate state for disinfection. CD is a sterilant and a true gas requiring a shorter contact and cycle times than VHP and does not exhibit the corrosive properties of liquid chlorine dioxide. In addition to their standard functions, both cage washers and autoclaves have been modified to serve as decontamination chambers for VHP and CD.

Waste containers should meet the following criteria:

- Easily available
- Sanitizable, leakproof plastic or metal construction with tight-fitting lids
- Have plastic liners available
- Labeled and/or easily distinguishable between trash, food, hazardous, and nonhazardous waste

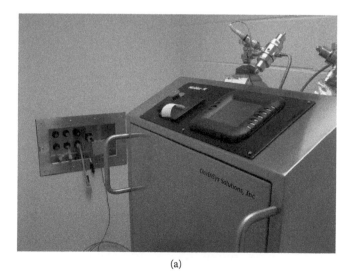

(a)

Fig. 26 (a) Room fumigation ports used in walls.

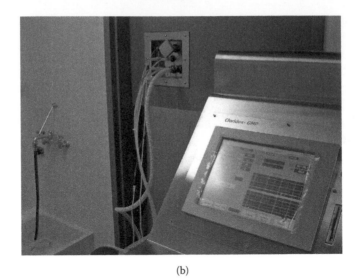

(b)

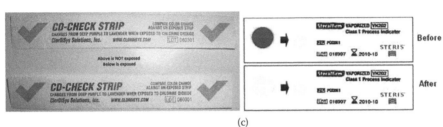

(c)

Fig. 26 (*Continued*) (b) Room fumigation ports used in doors. (c) Chemical indicators for gaseous chlorine dioxide and vaporized hydrogen peroxide (VHP). The biologic indicators for chlorine dioxide and VHP are *Bacillus atrophaeus* and *Geobacillus stearothermophilus*, respectively. (Images courtesy ClorDiSys Solutions Inc. and Steris Corporation.)

- Animal-carcass containers: leakproof disposable liners suitable for refrigeration/freezing
- Physically separate carcass and tissue storage from facilities housing animals or storing any materials that might contact live animals; maintain carcass storage temperatures < 7°C
- Frequently clean and disinfect storage areas and containers; keep storage areas free of insects and other vermin
- Handle hazardous waste according to local, state, and federal regulations

transportation

It is prudent to obtain information concerning the animals' genetic and microbiologic status before ordering from vendors (approved vendors), other institutions (unapproved vendors), and other facilities within one's institution. Evaluate the information for timeliness (how recent the information is), location (whether this information is from the same room/facility as the animals received), and other pertinent details. Animals from unapproved vendors or facilities with a history of disease outbreaks should undergo quarantine prior to admission to the general population.

Animal transportation, ranging from global to an adjacent room, exerts stress on the animals. The International Airline Transport Association (IATA), Laboratory Animal Breeders Association of Great Britain, Limited (LABA), Laboratory Animal Science Association (LASA), and the Institute for Laboratory Animal Resources (ILAR) have animal shipping recommendations. The goal of transportation is to minimize the animals' stress level by minimizing transit time and providing sources of food and water and adequate filtered ventilation. It is also important to avoid overcrowding, physical trauma, and environmental extremes. Ship only healthy animals and avoid shipping near-term pregnant animals, especially those within 5–7 days of parturition.

Disposable commercial transport enclosures meeting the food, water, and ventilation needs, while preventing cross contamination, are available. Pack animals in a HEPA-filtered environment and in a sound, previously autoclaved container, preferably with autoclaved bedding, and ensure that the container's integrity (including the filter material) is intact before removing the container from the HEPA environment. Add a water and food source and close the container appropriately. Avoid using produce (e.g., apples, potatoes) as a water source, especially for international shipment; instead use gel packs or liquid water. Distribute the cage's weight evenly to avoid handling mishaps; provide sufficient external warning in the case of unbalanced or particular heavy shipping containers. Securely affix the health records to the container's exterior and ensure that all local and international regulations are followed. A container top providing a visualization window is preferable. Fredenburg et al. (2009) evaluated various parameters following a 5-day ground transit. The use of gel resulted in higher food consumption and weight gain. Dehydration-associated blood parameters were more common with autoclaved potatoes than with gel. Elevations in ALP, ALT, phosphorus, and potassium were reported. The authors reported that lower shipping-container animal density resulted in higher food consumption and lower shipping-container temperatures.

When received, the container's outer surface has been subjected to contamination during shipment and therefore must be appropriately decontaminated before being brought into the animal facility/room and again immediately before opening. Several disinfectants are available, including a dilute bleach solution (28 mL/L of water). After an appropriate contact time, open the shipping container inside a HEPA-filtered (e.g. cage-change station) environment.

Provide an animal stabilization period (acclimation) after receiving animals from approved vendors. **Table 12** gives acclimation periods during which various research parameters revert (adapt) to pre-transportation values. Feirer et al. (2011) evaluated mouse altitude adaptation (transportation from a lower to higher altitude), finding that hematology and cytokine values were affected; the hematology data mimics changes observed by one of this text's authors (PES) when conducting rat-based research at altitude.

When considering international rat shipments, one should contact the responsible agency or agencies in the importing/exporting countries, including the appropriate individuals at the receiving institution. Animals may quickly pass through customs, to end up at a loading dock unbeknownst to those directly responsible for animal care. To minimize customs delays, one should discuss the shipment with the responsible authorities prior to requesting animals. For importation, these authorities would include the USDA (United States Department of Agriculture) and U.S. Customs; for exports, the USDA, the importing country's embassy, and individuals at the receiving institution.

TABLE 12: RAT TRANSPORTATION ACCLIMATION PERIODS BY
VARIOUS RESEARCH PARAMETERS

Research Parameter	Minimum Adaptation Period
Body weight	<24 hr
Body weight gain	2 d
Water intake	17–23 d
Luteinizing hormone (LH)	>7 d
Leukocytes (segmented neutrophils)	12 d
AST, LDH, potassium, cholesterol	12 d
Fecal corticosterone	6 d
Altitude (hematology/cytokine levels)	~21 d

quality control

Rat quality control concerns their microbiological and genetic background, and is an active process involving sentinel rats and serial serologic analysis for evidence of pathogens. Various microbes can have a negative impact on research projects and result in a loss of time, money, and rats. This is especially important since rats can be in a persistently infected carrier state with some pathogens, even after resolution of clinical signs. For these reasons it is important to receive animals free of known pathogens whenever possible and maintain them in a manner preventing infection. The animals received should be SPF (specific pathogen free) or free of viral antibodies. Maintenance may vary from conventional housing, IVCs, or the use of semirigid or flexible film isolators. Sentinel and quarantine programs are discussed in more detail in Chapter 4. Sanitation records, in addition to sentinel program records, are important components of the institutional quality control program.

Rat genetic integrity requires sound husbandry, personnel, and management practices. Assessment of a strain's genetic integrity reduces experimental variability. Methods used to assess genetic integrity include DNA-typing, morphometric, and biochemical methods. One may obtain this information from vendors or submit appropriate samples for microsatellite analysis.

identification and records

Accurate, concise research records necessitate animal identification. Some research areas may require meticulous record keeping, such as a drug or medical device GLP study. Without adequate identification (individual or group) it is difficult to maintain records. Individual rats may be identified by the following methods (listed least to most permanent):

- **Markers and dyes** are available for temporary identification on an animal's tail or fur. While this identification method appears benign, Burn et al. (2008) reported significantly elevated time in the open arms of the elevated plus maze and greater chromodacryorrhea when handled.

- Metal **ear tags** may be used, each with a unique number, to identify rats. These tags, however, even when properly placed, may be removed by the animal itself or its cagemate(s).

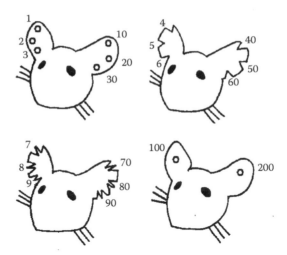

Fig. 27 Example of an ear punch/notch identification scheme.

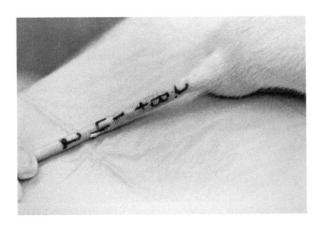

Fig. 28 Tail tattooing as a method of rat identification.

- **Ear punches (Figure 27)** are an easy and permanent means to identify animals. Identification depends on the location (left vs. right ear and location on a given ear) and number of notches. This system is highly variable, with individual laboratories developing their own identification systems.

- **Tattoos** are an effective and permanent means to identify rats **(Figure 28)**. The skill and experience of the individual tattooing the animal determines the tattoo's quality. Besides identification, tattoos are an effective means to identify boney prominences in kinesiology studies, tumor placement sites, etc. Tattooing digits ("microtattoo") in neonates is a humane alternative to digit amputation. Kasanen et al. (2011) found microtattooing caused less heart rate and blood pressure changes compared to ear tattoo and ear notching.

- **Electronic transponders** are also available for unique identification of animals. These small devices are implanted subcutaneously and are read with a transponder detector. Some transponder and detector models are programmable and permit temperature measurement (**Figure 29**). This method is

(a)

(b)   (c)

Fig. 29 (a) Electronic transponder readers. (b) Microchip transponder (actual size and high magnification view). (c) Microchip transponder in a trochar ready for implantation. (Image (a) courtesy of Bio Medic Data Systems, Inc.; images (b) and (c) courtesy Trovan, Ltd.)

very efficient but expensive. Another disadvantage is that the technology currently lacks standardization, requiring a lab to purchase the chips and detectors from the same manufacturer, as a manufacturer's chips typically cannot be read by another's detector.

Note: Digit amputation (toe clipping) should be discouraged as a method of identification; digit tattooing may be considered as a more humane alternative should identification of pups at an early age be necessary.

Usually, groups and individual animals are identified by cage cards. Cage cards should carry the following information:

- Species
- Stock or strain
- Sex
- Source/vendoProtocol number
- Investigator
- Investigator's department
- Other information: contact person, email, phone, etc.

In addition, individual animal records may be expected. Medical treatments, surgery, breeding programs, and studies involving hazardous materials are instances where individual records are essential.

3

management

overview

This chapter discusses programmatic management based on the ethical and responsible use of laboratory animals in research, teaching, and testing. Russell and Burch's *The Principles of Humane Experimental Technique* (1959) and their 3Rs (Reduction, Replacement, and Refinement) are important and frequently cited considerations. Besides applying the 3Rs, national animal welfare regulations frequently include husbandry and species-space requirements, veterinary care, ethics committee (e.g., Institutional Animal Care and Use Committee, or IACUC) requirements, research-related hazards (e.g., allergens, biohazard, toxins, carcinogens) and facilities/physical plant requirements.

This chapter provides a general overview, and readers are reminded that the regulatory landscape is frequently changing, as evident by the updated, 8th edition of the (U.S.) *Guide* and the Council of Europe's *European Treaty Series (ETS) 123*. In addition to national legislation there may be local and state/provincial legislation; furthermore, regulations applying to other, tangential uses of research animals (e.g., transportation, biohazards) will apply.

Legislation depends upon strong financial and administrative support by the Institutional Official (IO) for the animal care and use program. Effective animal care and use programs are inherently expensive and if not appropriately supported will frustrate the IO, Attending Veterinarian, IACUC/Ethics Committee, and, most unfortunately, the animal users. Effective care and use programs need the ability to support the institution with many individual animal users; large users should realize that "a high tide raises all boats," as there may be more pressing institutional needs. Ineffective animal programs will eventually "run adrift" institutionally and on regulatory compliance and voluntary accreditation, with the resulting correction costing more than regular, consistent attention to this important and essential research support group.

Along with the 3Rs and national legislation, there are two recognized international bodies with guidelines on the ethical use of animals in biomedical research.

Council for International Organizations of Medical Sciences (CIOMS)

CIOMS was established by the World Health Organization (WHO) and the United Nations Educational, Scientific, and Cultural Organization (UNESCO) in 1949. CIOMS develops wide-ranging programs on various topics, including bioethics. The *International Guiding Principles for Biomedical Research Involving Animals* (1985) outlines 11 key ethical expectations for animal-model-based biomedical research programs. The revised *Guiding Principles* is expected to be released in 2012.

The U.S. Office of Laboratory Animal Welfare (OLAW) uses the CIOMS *Guiding Principles*, in part, when foreign institutions apply for an Animal Welfare Assurance. According to OLAW, "Submitting a Foreign Assurance for OLAW approval commits the institution to follow the *International Guiding Principles for Biomedical Research Involving Animals* developed by the Council for International Organizations of Medical Sciences (CIOMS) and to comply with all applicable laws, regulations and policies governing animal care and use for their country of jurisdiction."

Office International des Epizooties (OIE)

The OIE was founded in 1924 to fight animal disease, maintaining the original acronym it became the World Organisation for Animal Health in 2004. The organization's *Terrestrial Animal Health Code* (2010) includes the chapter "Use of Animals in Research and Education." The document's intent is "to provide advice and assistance for OIE Members to follow when formulating regulatory requirements, or other form of oversight, for the use of live animals in research and education. A system of animal use oversight should be implemented in each country. The system will, in practice, vary from country to country and according to cultural, economic, religious and social factors. However, the OIE recommends that Members address all the essential elements identified in this chapter in formulating a regulatory framework that is appropriate to their local conditions."

select regulatory agencies and compliance

Australia

The *Australian Code of Practice for the Care and Use of Animals for Scientific Purposes* (2004) is a document published by the Australian National Health and Medical Research Council (NHMRC). The *Code of Practice* was originally published in 1969. The *Code of Practice* was written and reviewed by the NHMRC, the Commonwealth Scientific and Industrial Research Organization (CSIRO), the Primary Industries Ministerial Council, the Australian Research Council, the Australian Vice-Chancellors' Committee, Australia State and Territory Government representatives, and animal welfare organization representatives. Although not legislation, researchers receiving funds from the above sources are expected to follow the *Code*. Some states have additional legislation and have incorporated the *Code* into their regulations. The *Code of Practice* is currently undergoing revision (due out in 2012).

Canada

Created in 1968, the Canadian Council on Animal Care (CCAC) sets forth the control of the scientific use of vertebrates and cephalopods in research, teaching, and testing as an autonomous and independent organization. The CCAC establishes policies, guidelines, and, in turn, standards that apply nationwide. In addition, some provinces have regulations concerning the use of laboratory animals. The CCAC certifies institutional compliance and assesses institutions triennially in a manner similar to the Association for the Assessment and Accreditation of Laboratory Animal Care (AAALAC-International).

Europe

Regulating animal use for experimental and scientific purposes occurs at the provincial/state, national, and international levels. The two international organizations responsible for regulating animal use are the 47-member Council of Europe (*Convention for the Protection of Vertebrate Animals Used for Experimental and Other Scientific Purposes*) and the 27-member European Union (The European Union's Directive 2010/63, *Protection of Animals Used for Scientific Purposes*). The 27 European Union (EU) states are also members of the Council of Europe (CE). Legislative enforcement is the individual country's responsibility. Neither the CE Convention nor the EU Directive prevents members from adopting stricter legislation. With a few exceptions, there is a high level of agreement between the two policies. The CE countries enact regulation meeting the minimum standards set forth in *ETS 123*.

People's Republic China

The Ministry of Science and Technology (MOST) has primary administrative oversight of animals used in research, teaching, and testing. Various autonomous regions, provinces, and municipalities have developed their own laws, regulations, and standards, although centered on the MOST's *Guidelines on the Humane Treatment of Laboratory Animals*, established in 2006. A national licensing and registration system exists for laboratory animal breeders and users that is managed by the MOST. The MOST provides guidelines on animal import, export, and transportation, bedding, drinking water, and other standards.

Taiwan, Republic of China

The Taiwan Animal Protection Law (1998) covers animals that are "fed or kept for the purpose of scientific application." The law is implemented by the Council of Agriculture (COA). The COA's Experimental Animal Ethics Committee conducts periodic institutional inspections.

India

India's Prevention of Cruelty to Animals Act (1960) covers "Experimentation of Animals." The Act exempts the states of Jammu and Kashmir, establishes the Animal Board of India, and permits the use of animals for experimental purpose for the benefits of humans, animals, and plants. The Act discusses the Central Government's ability to establish a "Committee for control and supervision of experiments on animals ... for the purpose of

Act (PL 99-158) provisions. This includes awards involving the use of rats. The standards are those described in the current *Guide*. OLAW is currently responsible for enforcing the Act. Impacted institutions must prepare and negotiate an Assurance statement that varies depending on the institution's AAALAC accreditation status; foreign institutions receiving such awards must also file an Assurance. An acceptable Assurance is valid for up to 5 years.

The U.S. Good Laboratory Practice (GLP) Act contains regulations for conducting nonclinical laboratory safety studies for products regulated by the Food and Drug Administration (FDA). The regulations contain requirements for animal care (including rats), animal facilities, and study records.

The IACUC is the basic unit of an effective animal care and use program. The PHS (when PHS funds are being used) requires an IACUC at any institution using rats in research, teaching, and testing. Important points regarding the IACUC's composition include the following:

- PHS policy requires a minimum of five members, including:
  - A chairperson.
  - A doctor of veterinary medicine who has training or experience in laboratory animal medicine or science, and responsibility for activities involving animals at the research facility.
  - An individual who is in no way affiliated with the institution other than as an IACUC member; some institutions use clergy, lawyers, or local humane society or animal shelter officials.
  - A practicing scientist with experience in animal research.
  - One member whose primary concerns are in a nonscientific area. This individual may be an employee of the institution served by the IACUC.
  - One individual can fulfill more than one of the above categories.
- IACUC responsibilities:
  - Protocol review for activities involving animals in research, teaching, and/or testing. Protocol approval by the IACUC must occur before animal use can begin.
  - Inspect and ensure the animal research facilities and equipment meet acceptable standards.
  - Ensure personnel are adequately trained and qualified to conduct research using animals.

- Ensure adequate handling and care for animals.
- Ensure considerations for alternatives to potentially painful and stressful procedures and determine the research is nonduplicative.
- Ensure appropriate use of sedatives, analgesics, and anesthetics.
- Ensure proper surgical preparation and techniques are used.
- Ensure appropriate euthanasia techniques are used.

Table 13 summarizes animal welfare guidelines and legislation.

voluntary accreditation

AAALAC-International

The Association for Assessment and Accreditation of Laboratory Animal Care-International (AAALAC-International) is a nonprofit organization providing a voluntary, peer-review-based accreditation to institutions using animals in research, teaching, and testing. AAALAC, established in 1965, is perceived as the "gold standard of laboratory animal care."

AAALAC's three primary regulatory standards are the *Guide*, the *Guide for the Care and Use of Agricultural Animals in Research and Teaching* (*Ag Guide*, not applicable to rats), and *ETS 123*, Appendices A and B. For U.S. institutions, the current version of the *Guide* along with other pertinent regulations apply. A given nation's rules and regulation apply to foreign institutions; should there be a foreign regulatory "void" (e.g., occupational health and safety), the *Guide* or associated regulation applies, and in the case of a regulatory conflict, the stricter regulation applies.

AAALAC's accreditation process requires an institution to submit a written (in English) Program Description (PD) highlighting institutional compliance with the pertinent regulatory standards. The PD format follows the *Guide*'s subject headings and may be downloaded from AAALAC's website. The PD is reviewed by the AAALAC site visitors before the site visit. The site visit is generally conducted with at least two site visitors, one being an AAALAC council member and the other(s) being AAALAC ad hoc specialists/consultants. Pre-site visit communications are coordinated through the Council member and

TABLE 13: SELECT NATIONAL EXPERIMENTAL-ANIMAL WELFARE GUIDANCE DOCUMENTS AND/OR LEGISLATION

Country	Document (Year)
Australia	Australian Code of Practice for the Care and Use of Animals for Scientific Purposes (2004)
Austria	[Bundesgesetz über den Schutz der Tiere (Tierschutzgesetz – TSchG)] Federal Act on the Protection of Animals (2004)
Chile	[Sobre Protección de Animales] Animal Protection Law (2009)
Croatia	The Animal Protection Act (2006)
Cyprus	The Bioethics Law of 2001 (2001)
Czech Republic	Animal Protection Act 283/1992
Estonia	Animal Protection Act (2000, 2002) Government Regulation No. 187 (2003)
Finland	Animal Welfare Act (2006)
Germany	[Tierschutzgesetz] Animal Welfare Law (2010)
India	Prevention of Cruelty to Animals Act (1960) Animal Welfare Act (draft, 2011)
Kenya	Prevention of Cruelty to Animals Act (1983)
Liechtenstein	Total ban on animal use in experimental procedures
Malaysia	Animals Act (1953), Amendments (2006), The Revision of Laws Order (2006)
New Zealand	Animal Welfare Act (1999)
South Africa	The Care and Use of Animals for Scientific Purposes (2008) Veterinary and Para-veterinary Professions Act (1982)
Norway	Norwegian Animal Welfare Act (2010)
Peru	Protection of Domestic Animals (2009)
Philippines	Act to Promote Animal Welfare in the Philippines (1998)
Poland	Polish Animal Protection Act (1998)
Romania	Law No. 205/2004 on the protection of animals (2004)
South Korea	Korean Animal Protection Act (2011)
Sweden	[Svensk författningssamling djurskyddslagen] [Svensk författningssamling djurskyddsförordningen] Swedish Animal Welfare Act (2009: 303) Swedish Animal Welfare Ordinance (2009)
Switzerland	Swiss Federal Act on Animal Protection (1978)
Taiwan, ROC	Taiwan Animal Protection Law (1998)
Tanzania	Animal Welfare Act (2008)
Zimbabwe	Prevention of Cruelty to Animals Act (1986) Scientific Experiments on Animals Act (1963)

a designated institutional representative. The site visit length and the number of site visitors vary with institutional size and complexity. The site visit involves an Entrance Briefing, a Program Review (including a PD review and clarification), meetings with various institutional representatives (e.g., IO, IACUC committee/chair, attending veterinarian), document review (e.g., research protocols, compliance records), and facility review (e.g., animal housing, research areas).

At the end of the site visit, site visitors enter into an Executive Session discussing, among themselves, their site visit findings. After the Executive Session the site visitors, usually led by the Council Member, conduct an Exit Briefing outlining their findings with institutional representatives. Findings consist of "mandatory items" or "suggestions for improvement." At this time, site visitors can convey their recommendation to the AAALAC council. Institutions have 10 days to respond to the site visitors' findings in a Post Site Visit Communication (PSVC). The site visitors' findings, with the PSVC, will be presented at the next AAALAC council meeting, with the site-visit coordinating Council member serving as the institution's advocate. Council meeting results (**Tables 14** and **15**) are conveyed,

TABLE 14: POSSIBLE **AAALAC** ACCREDITATION CATEGORIES FOR NEW APPLICATION INSTITUTIONS

Accreditation Category	Considered Accredited?	Mandatory Item Correction Time
Full	Yes	N/A
Conditional	Yes	Correct mandatory item(s) by the AAALAC annual report or as determined by AAALAC Council
Provisional	No	Up to 24 mo
Withhold	No	N/A

TABLE 15: POSSIBLE **AAALAC** ACCREDITATION CATEGORIES FOR EXISTING ACCREDITED INSTITUTIONS

Accreditation Category	Considered Accredited?	Mandatory Item Correction Time
Full	Yes	N/A
Conditional	Yes	Correct mandatory item(s) by the next AAALAC Annual Report or at the discretion of AAALAC Council
Deferred		Up to 2 mo; uncorrected mandatory item(s) will result in Probation
Probation	No	Up to 12 mo; uncorrected mandatory item(s) will result in Revoke
Revoke	No	N/A

in writing, most often to the IO. Generally, negative letters are received shortly following the council meeting, while positive results are received after a longer interval.

zoonotic diseases/occupational health

Zoonoses are diseases transmitted from animals to humans. Infectious agents can be transmitted by aerosol, ingestion, skin wounds, and mucous membrane exposure (e.g., conjunctiva). Protective clothing, biosafety cabinets, specialized housing, and shower facilities may all be required depending on the biohazard. Specific procedures and facility requirements for safely working with various biohazard agents are discussed elsewhere (e.g., the Center for Disease Control and Prevention's (CDC's) *Biosafety in Microbiological and Biomedical Laboratories*, or *BMBL*). Institutions using these agents must have appropriately trained environmental health and safety (EHS) and occupational health and safety programs (OHSP) staffed by knowledgeable professionals to provide the IACUC and animal users with information, guidance, and assistance. OHSP directors should also consider the possibility that individuals conducting field studies may be exposed to plague (*Yersinia pestis*), tularemia (*Francisella tularensis*), and other atypical agents endemic in certain areas of the world.

It is important for individuals enrolled in an institutional OHSP to accurately convey their medical history and understand it will be handled confidentially. While this may not be universal outside the United States, institutions must develop and maintain such a confidential process; otherwise, the information collected in the medical history questionnaire may be unreliable or may unnecessarily jeopardize employee health and safety. Enrollment in an institutional OHSP should not be restricted to those with direct animal contact; it should include janitorial and physical plant staff, visiting scientists, and other individuals with vivarium access. Researchers removing animals from the vivarium for further study might be sharing laboratory space with others, and this may have consequences; all such individuals must be included in the OHSP. Some programs may require serum banking.

Equipment and PPE requirements should be based on a project's risk assessment and the contribution of knowledgeable EHS personnel; equipment should complement the research program and will benefit from researcher input. Task-appropriate equipment such as down- and back-draft tables, BSCs, ventilated storage cabinets, and snorkels enhance personnel safety. Workers should wear gloves, a mask, and a dedicated laboratory coat when handling rats. At a minimum, gloves, an appropriate mask, a laboratory coat, and protective eyewear should be worn for any necropsy or while working with fresh or fixed tissue or body fluids from any animal. Thorough hand washing before and after any procedure using animals is one of the best preventative measures.

Zoonotic Diseases

Select zoonotic diseases associated with rats include the following:

Hantaviruses

Rats, including the cotton rat (*Sigmodon hispidus*) and the rice rat (*Oryzomys palustris*), are susceptible to hantaviruses, enveloped viruses of the family Bunyaviridae. These viruses cause the human diseases hantavirus pulmonary syndrome (HPS) and hemorrhagic fever and renal syndrome (HFRS). Hantaan virus, a member of the HFRS group, produces no clinical disease in rats. HFRS in humans is characterized by fever, thrombocytopenia, and capillary leakage leading to retro-peritoneal and renal hemorrhage. The capillary leakage in HPS is localized in the lungs.

Cowpox/ratpox

Cowpox virus (CPXV), an *Orthopoxvirus* related to vaccinia virus (VV), reportedly occurs in rats from parts of Europe and is an important emerging zoonosis. Wild rodents are considered the true reservoirs, but CPXV's prevalence in rats is unknown. Cattle, cats, zoo animals, and humans are incidental hosts. Small outbreaks and rare human CPXV infections occur by rat bite or direct skin contact with infected rats. Person-to-person transmission has not been reported. In humans, CPXV usually causes localized skin lesions, including ulcerated nodules and molluscum-like lesions; however, human clinical disease depends on the patient's immune status and VV vaccination. Immunocompromised patients develop severe generalized skin infections; VV-vaccinated patients have less severe symptoms and delayed disease onset. European wild rats were responsible for CPXV clinical disease in Barbary macaques (*Macaca sylvanus*) and an elephant (*Elephas maximus*). CPXV-infected rats may be asymptomatic, possess distinct skin lesions on extremities, mouth, and nose, and/or die. Rats may be tested by IFA or PCR/virus isolation from skin lesions and brain.

A cowpox-related virus has also been reported in laboratory rats from Europe and the former Soviet Union. Referred to as Turkmenia rodent poxvirus, it is distinct from but has similar clinical signs to mouse ectromelia virus. The virus causes silent infections, dermal pox, tail amputation with high mortality, and interstitial pneumonia. Whether the Turkmenia rodent poxvirus is, in fact, cowpox virus is not known.

Lymphocytic choriomeningitis virus (LCMV)

This enveloped member of the family Arenaviridae is of primary importance as a zoonotic agent and as a contaminant of transplantable tumors and cultured cell lines. Most rodents are susceptible to infection, with hamster infections being especially common and rats being naturally resistant. However, neonatal Lewis rats experimentally inoculated with LCMV developed distinct retinal, cerebellar, and hippocampal neuropathology. Disease is generally seen only following experimental intracerebral inoculation and is not a feature of natural infection. Serology is usually included in routine colony health screening tests.

Streptobacillus moniliformis (rat-bite fever, or haverhill fever)

This Gram-negative, pleomorphic rod or filamentous organism causes the zoonotic disease rat-bite fever. The bacteria grow slowly (requiring up to seven days) and require trypticase soy agar or broth enriched with 20% blood, serum, or ascitic fluid. S. *moniliformis* is currently uncommon in laboratory rats but is common in wild rats. It is associated with opportunistic respiratory infections and can cause wound infections and abscesses. In humans, the incubation period is less than 10 days, causing a rash frequently involving the extremities, including the palms and soles (**Figure 30**), fever, headache, polyarthritis, and even death. A similar syndrome, Haverhill fever, has been associated with ingestion of rat-contaminated food, particularly milk. Elliott (2007) has an excellent S. *moniliformis* review. *Spirillum muris*, another commensal organism, has also been associated with rat-bite fever. Unlike S. *moniliformis*, arthritis is rare and the S. *muris* (or *sodoku*) human incubation period is longer (1–6 weeks).

Leptospirosis

Common *Leptospira* serovars (bacterial species or subspecies variants) in animals include *canicola, hardjo, autumnalis, icterohemorrhagica, grippotyphosa*, and *pomona*. Leptospirosis is caused by a Gram-negative spirochete.

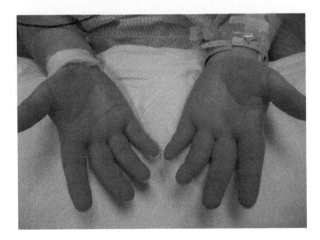

Fig. 30 Characteristic rash associated with *Streptobacillus monili-formis* infection in humans. This macular rash may occur on the palms or soles. (Image courtesy Lito Papanicolas.)

- Rodents can remain inapparent carriers for life.
- Human symptoms include fever, headache, myalgia, nausea, jaundice, stiff neck, chills, rash, and conjunctivitis.
- Transmission is by contact with urine-contaminated water or direct contact. In the United States the most important sources of infection are from dogs, farm animals, wildlife (especially rodents), and cats. However, human disease is associated with laboratory and pet rodents.
- Considered an occupational disease of veterinarians, farmers, military personnel, and abattoir workers.

Salmonellosis

Salmonellosis, caused by a Gram-negative bacteria, is relatively rare in well-managed rodent facilities.

- Clinical signs in the rat include anorexia, decreased activity, rough hair coat, and soft and formless feces. Animals can be convalescent carriers or active shedders, or act as fomites. Infection can result in the animal's death.
- In humans, clinical signs include abdominal pain, diarrhea, fever, and vomiting.
- Transmission is fecal-oral.

Plague

Plague (black plague, bubonic plague) is caused by the Gram-negative bacterial rod *Yersinia pestis*. It was responsible for the death of up to one-third of Europe's population in the 14th century. The organism is most frequently transmitted by fleas. Individuals performing field studies, even while working on other species (e.g., birds, bats) may inadvertently contact rats or infected fleas. The disease is endemic in the southwestern United States and other parts of the world.

Cestodiasis

Hymenolepis nana occurs in several rodent species. Unlike other cestodes, where an insect intermediate host is needed, direct transmission is possible.

- Rat infection may be asymptomatic or cause intestinal obstruction and enteritis.
- Human clinical signs include enteritis, anorexia, anal pruritus, and headache.
- Transmission is fecal-oral via ingestion of feces from infected animals.

Acariasis

Ornithonyssus bacoti (**Figure 31**), the tropical rat mite, and *Laelaps echidninus*, the spiny rat mite, are rarely encountered. Both mites remain on the host only long enough to obtain a blood meal. They are not host-specific, will bite humans, and can act as vectors for several agents pathogenic to humans. *O. bacoti* is the most commonly reported mesostigmatid mite both in laboratory rodent colonies and in human-associated rat mite dermatitis cases. The mites are on the host only when feeding, after which they seek refuge in the surrounding environment. They can cause anemia, debility, and infertility in rats. Besides causing dermatitis by the mites feeding, they can experimentally spread a host of human agents, including Q fever, plague, and Eastern and Western encephalitis.

Dermatophytosis (ringworm)

This superficial fungal infection involves the keratinized layers of the skin and its appendages.

- The most common rodent dermatophyte is *Trichophyton mentagrophytes*. Rat skin lesions are characterized by alopecia, broken hairs, erythema, or crusts. Uncommon in laboratory rats.

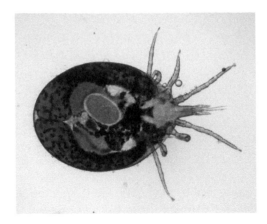

Fig. 31 *Ornithonyssus bacoti* (tropical rat mite). Female mites are normally ~750 µm in length, but can increase to greater than 1 mm when engorged. (Image courtesy Lisa Heath.)

- Human lesions include papulosquamous (scaly, red, elevated skin areas), dry, with a red ring (erythema) affecting any part of the body.
- Transmission is through physical contact. Several human dermatophytes can cause ringworm (anthropophilic); transmission is from human to human, and they are not associated with animals.

Bite or scratch wounds

Besides rat-bite fever, rat bites and scratches, especially as improperly treated wounds, can become infected with a variety of bacteria, including those from the rats' environment. Deep puncture wounds provide an entry for tetanus organisms, and therefore tetanus immunization is highly recommended for those working with animals, including rats. All institutions should have an easily accessible bite/scratch kit and an institutional policy for personnel receiving such work-related wounds.

Occupational Health

Allergies

Human hypersensitivity reactions to rat allergens (e.g., urine, dander) are historically relatively common. The primary rat allergens are the urinary proteins Rat n1, with Rat n2 having a lesser role; fur and epithelia may be responsible, in part, from urine contamination. Symptoms vary from hay-fever-like conditions to asthma, skin wheals, and eczema. Allergic sensitivity usually develops within two years of working with animals. Individuals with a history of allergies may develop more serious manifestations to animal allergens (e.g., asthma). Often a variety of laboratory and clinical tests, including skin tests for animal dander, are needed to establish a diagnosis. Sensitization to rat and mouse (and possibly other animals) urinary allergens likely results from cross-reactive molecules rather than atopy. The Rat n1 and the Mus m1 lipocalin proteins share similar 3D structures, cross-reactive IgE binding determinants, and a 66% amino acid sequence homology. The lipocalin family may play a role with rodent pheromones, and, in the case of Rat n1, has a similar response to the cow lipocalin allergen Bos d2. Male rodents may excrete up to 100 times the urinary allergens than females. While no relationship was found between rat allergy symptoms and endotoxin levels, those who smoked cigarettes were more likely to develop symptoms without IgE sensitization. Protection against developing animal allergen allergic reactions includes using proper equipment (e.g., microisolator tops, IVCs, cage change stations), proper PPE (e.g., masks, fitted respirator, gowns, dedicated clothing), and showering after the workday. Higher Rat n1 concentrations are associated with procedures involving feeding, cage cleaning, handling, and injection. There are no occupational health limits for rodent allergens (e.g., rat/ mouse urinary protein); however, institutions should consider work-practice monitoring (e.g., lateral flow immunoassays during cage changing, bedding disposal) and ensuring that employee medical history questionnaires are periodically and accurately updated and reviewed. Krop et al. (2007) determined that hair-covering allergen levels correlated with airborne allergen levels. Eggleston et al. (1990) suggested an allergen concentrations <5 ng/m^3, although this level failed to prevent allergic symptoms. Multiplex assays for indoor allergens have been described and may prove beneficial for work-practice monitoring. In addition to rat (and other animal) allergies, individuals may develop allergies to materials such as latex.

Ergonomics

Ergonomics, defined by the International Ergonomics Association (IEA), is a "scientific discipline concerned with the understanding of the interactions among humans and other elements of a system, and the profession that applies theoretical principles, data and methods to design in order to optimize human well being and overall system performance." In the vivarium this involves various systems, including proper computer use, equipment interaction (e.g., cage and water bottle processing), and lifting techniques to minimize musculoskeletal disorders (MSD). According to 2008 U.S. Bureau of Labor Statistics, individuals with MSDs were away from work 10 days; greater than 60% of all employees with MSDs were away from work for 6 or more days. This is a significant financial burden for employers and a negative impact on employee psychology and productivity. Besides physical ergonomics, the IEA also concerns itself with cognitive and organizational ergonomics.

By working with institutional occupation health and safety professionals, vivarium personnel at all levels can learn to develop better work practices to minimize the occurrence of MSDs. Measured solutions complement the institutional training and awareness. Solutions include providing more ergonomic handles and grips, self-leveling racks to handle cages and water bottles more correctly, and process automation (e.g., robotics, automatic watering). Ergonomic solutions warrant strong consideration during facility renovation and new construction.

Bloodborne pathogens

The U.S. Department of Labor's Occupational Safety and Health Administration (OSHA) has developed bloodborne-pathogen regulations covering "occupational exposure to blood and other potentially infectious materials." Other potentially infectious materials (OPIMs) include various body fluids, unfixed tissues (excluding intact skin), and human immunodeficiency/hepatitis B virus–contaminated/containing cells and tissues. Most facilities include animals with tissue exposure (e.g., rats receiving human breast cancer cell xenoplants), even if the tissues were deemed negative based on human-pathogen PCR-based testing (e.g., h-IMPACT from IDEXX RADIL™). Vaccinations are offered to employees working with hazards such as hepatitis B.

Individuals working with such animals are expected to receive bloodborne pathogen training initially (and then annually), and they are to exercise "universal precautions" to prevent exposure. Universal precautions include all blood and OPIMs being handled as if infected, including cages housing exposed animals; proper PPE being worn; and great caution being exercised when working around sharps. Bloodborne pathogen exposure may occur by skin (e.g., bite, scratch, or existing skin wound), mucous membranes, inhalation, or inoculation (e.g., needlestick).

During training, workers must be instructed how to follow the accepted institutional practices set forth by the institution's occupational health/biosafety professionals when an exposure occurs. Postexposure activities frequently include washing (with a disinfectant) or flushing (e.g., eyes) the exposed area and reporting to the institutional occupational health facilities or emergency room depending on the severity of the exposure and whether the exposure occurred after work hours or on a weekend or holiday. Proper and timely exposure documentation is required.

veterinary care

Good, adequate veterinary care for laboratory animals is a moral, legal, and scientific responsibility. For laboratory rodents, the veterinary care program can be directed to an individual animal or to the entire colony. It not only involves the diagnostic assessment of a sick animal, being composed of multiple components (e.g., quarantine and sentinel programs) serving the general and ultimate purpose of maintaining the entire colony's health and well-being. When a single animal is sick, it is best to identify other colony animals manifesting similar clinical findings and establish a prevalence rate for the diagnosed disease. More often than not, a factor common to the sick animals in a group can be established. This can help identify the cause of the problem and prevent additional morbidity and mortality.

clinical history

Signalment and clinical/experimental history are a fundamental component of a diagnostic assessment. Basic data, including an animal's age, sex, and strain/stock, can help identify a problem list. For example, aging SD rats are prone to chronic progressive nephropathy and aging F344 rats tend to develop large granular lymphocytic leukemia. Some conditions can be incidental findings affecting certain strains/stocks of rats, while some phenotypes can be clinically manifested, too.

Obtaining a research animal's clinical history entails a thorough and detailed account of the animal's experimental purpose and an understanding of the experimental methodologies performed on the animal. One may expect laboratory animals to show clinical signs associated with the research; however, some studies (e.g., drug discovery, infectious disease) can have unpredictable clinical outcomes that necessitate the implementation of experimental endpoints and procedural refinements. Husbandry and environmental conditions also require evaluation. Examples in rats include low humidity and poor bedding quality associated with ringtail and blepharitis, respectively.

physical examination

A systemic approach for physical examination starts by carefully observing at a distance the animal in its home cage. The observer should note the rat's behavior, activity, and general appearance, including but not limited to respiratory pattern, presence of ocular and/or nasal discharge, and other signs of pain and distress (e.g., hunched posture and piloerection). The cage requires careful examination to determine whether the rat has been eating, drinking, urinating, and defecating normally.

One must be able to distinguish a morbid rat from a normal one. Rats are very stoic animals and may be quite ill before they actually show clinical signs of disease. The examiner should slowly approach the rat and note any depression of response to external stimuli. Finally, a thorough physical exam can be performed with the aid of a stethoscope, thermometer, and a tongue depressor/small speculum/ otoscope to aid in visualization of the teeth and a scale to measure the animal's weight in grams. The rat physical exam should include the following:

Vital Signs

Rats have a normal body temperature of 35.9–37.5°C (96.6–99.5°F). One may determine this parameter either rectally or by the use of an over-the-counter tympanic membrane thermometer. Thermistor probes are very accurate and are available for rats. Some microchips used for identification are capable of temperature determination. Whatever method is employed, it is important to note that handling may cause significant stress to the rat and elevate its body temperature (stress-induced hyperthermia). Thus, it is advisable to determine the body temperature first, before any other procedures are performed.

One may auscultate the heart and lungs with a pediatric stethoscope to identify abnormal heart/lung sounds and patterns. Cardiovascular and respiratory physiology can also be evaluated with telemetry devices. Rats are susceptible to a variety of respiratory pathogens described later in this chapter.

The patient's body weight (BW) should be taken. BW varies with an animal's age, sex, and stock/strain. Weight loss, measured as a percentage of the animal's initial BW or as compared with age-matched conspecifics, is often an indicator of a disease process. Yet weight loss may be associated with the research paradigm and not exactly reflect a poor quality of life. For example, diabetic rats typically have reduced muscle mass but still are clinically "normal." Also, studies creating physiological changes, such as intraperitoneal fluid retention or tumor growth, may mask weight loss by interfering with identifying reduced fat stores and muscle mass. Thus, a body condition scoring system, such as the one described by Hickman and Swan (2010), may more accurately reflect a rat's condition and nutritional state.

Organ System Evaluation

Eyes

Discharges and abnormalities of the rat's eyes and its related structures should be noted. Chromodacryorrhea, the secretion of "red tears" (porphyrin pigment), is a stress indicator, which may result from experimental manipulation or a from disease process, such as a coronavirus infection. Although there may be no pigment present around the eyes or face, one should check the forepaws for pigment, as rats will wipe the pigment away from the face during grooming. One may distinguish the pigment from blood by using a Wood's lamp, as porphyrin fluoresces and blood does not.

An ophthalmic exam is a component of systemic toxicology studies, and the review articles by Kuiper et al. (1997), Williams (2002), and Wegener et al. (2002) would be of interest to anyone performing such examinations. These sources provide an overview of testing procedures, equipment, international guidelines, and ophthalmic terminology.

Ears

Check the ears for any abnormalities, including any swelling or hair loss near the ears' base that may indicate Zymbal's gland tumor. Auricular chondritis, an autoimmune response characterized by bilateral auricular thickening, is associated with metallic ear tags.

Mucous membranes

Examination of the mucous membranes' (MM) color, moisture, and capillary refill time can help determine the rat's hydration status and clinical conditions such as acute cardiovascular compromise (cyanotic or pale MMs) and shock (hyperemic or muddy MMs).

Oral cavity

The rat's oral cavity is difficult to examine because of its small opening and the tendency of the buccal mucosa to encroach toward the middle of the mouth. Rats must gnaw to erode the continuously growing incisors. Overgrowth will occur in instances of tooth breakage of the opposite arcade, powdered/liquid diet studies, or malocclusion. Malocclusion is a heritable trait; affected animals should therefore not be bred. Teeth trimming is palliative and should be carefully performed to prevent shattering the tooth being trimmed. One may use a Dremel®, or other such product, to provide optimal control when teeth trimming.

Skin and hair coat

Poor hair coat usually results from lack of grooming and may indicate disease, pain and distress, cold, or poor nutrition. Rats are also susceptible to ectoparasites, which may also result in a poor hair coat and itching. Skin abnormalities such as alopecia should also be noted. Fecal staining and bedding material stuck on the perineum are indicative of diarrhea.

Lumps and bumps

Generally, masses may indicate an abscess, tumor, granuloma, or other conditions such as hematoma.

Abdominal palpation

Assess abdominal symmetry and gently evaluate the abdomen for the presence of masses, peritoneal fluid, rigidity, or unusual pain. Palpate the visceral organs including the kidneys, liver, spleen, and urinary bladder. Abdominal distention can be caused by organomegaly, ascites, *Clostridium piliforme* infection (Tyzzer's disease), intestinal distension associated with chloral hydrate toxicity (ileus), and abdominal mass, among others.

Genital and perineal region

Aggressive behavior, although less common in rats than in mice, is typically directed to the lower back, rear, and tail (**Figure 32**).

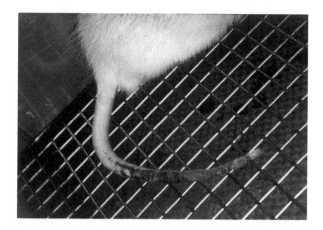

Fig. 32 Tail fight wounds. (Image courtesy Department of Comparative Medicine, College of Medicine, Penn State Hershey Medical Center.)

Another common finding is preputial gland lesions that include infection (abscess) and carcinoma/adenoma. Fecal and urine staining should also be noted.

Extremities

Note the presence of pain, heat, swelling, deformities, or range-of-motion limitations. Pododermatitis is a common problem of the extremities and associated with housing on wire-mesh cages longer than one year (Peace 2001), obesity, and trauma. Limb injuries can result from improper animal handling.

diagnostic imaging

Diagnostic imaging, or bioimaging, permits serial noninvasive viewing of tissues and organs and is available in many modalities, including radiographs, ultrasound, magnetic resonance imaging (MRI), computed tomography (CT), dual-emission x-ray absorptiometry (DEXA), positron emission tomography (PET), and single photon emission computed tomography (SPECT). The techniques can be used either for diagnostic purposes or as part of a research study, the latter being more common regarding laboratory rodents. In this section, we discuss the two most common methods of diagnostic imaging performed on rats. A more thorough discussion of bioimaging, including other imaging modalities, is found in Chapter 5.

Radiographs

Radiographic equipment capable of generating 300 mA, an exposure of 1/120 (0.008) sec, and 50–60 kVp is considered best for use in rats. Rats should be radiographed using a tabletop technique. Some groups have found the combination of conventional radiographic and mammography techniques useful in imaging tumor-bearing animals. Dental x-ray units that can focus at short distances can also be used for focal anatomy or full body radiographs.

Ultrasound

Ultrasound is another valuable diagnostic tool that in some cases may yield more information than radiographs. For best results use high-frequency transducers (7.5 MHz) with narrow sector angles.

There are also several different ultrasound modes. A-mode is the simplest, wherein a single transducer scans a line through the body with the echoes plotted on screen as a function of focal depth. In B-mode ultrasound, a linear array of transducers simultaneously scans a plane through the body that is viewed as a two-dimensional image on a screen. In M (motion)-mode, a rapid sequence of B-mode scans and the images follow each other in sequence for visualization and measurement. Lastly, Doppler mode uses the Doppler effect in measuring and visualizing blood flow. Additional techniques are also available and have been used on rats, including contrast-enhanced ultrasound (CEUS), where microbubble contrast agents enhance the ultrasound waves, resulting in increased contrast.

diseases of rats

Adventitious rodent infections, from both well-documented and emerging agents, continue to pose significant risk to biomedical research in spite of improved veterinary and husbandry practices and advances in engineering standards of laboratory rodent housing and handling. Concurrent rodent colony infections can produce subtle to devastating effects, which at best confound and at worst invalidate research data. For example, Kilham rat virus, *Mycoplasma pulmonis*, and Sendai virus can contaminate cell cultures and transplantable tumors and cause immune response alterations in rats. The *ILAR Journal* issue *Detection and Management of Microbial Contamination in Laboratory Rodents* (2008: 49(3)) provides excellent reviews on rodent infectious diseases.

This section is divided into discussions of specific disease conditions with descriptions of clinical signs and lesions, a discussion of treatment and control measures, and programs designed to prevent the introduction of infectious disease into the colony. Zoonotic diseases are presented in Chapter 3.

bacterial agents

Mycoplasmosis

This small, pleomorphic extracellular parasite lacking a cell wall colonizes the epithelial surface of the respiratory tract, middle ear, nasopharynx, and endometrium. It causes murine respiratory mycoplasmosis (MRM), formerly called chronic respiratory disease (CRD). Although rare in modern laboratory animal populations, it is common in pet and wild rat and mouse populations. Lesions may be acute or chronic, consisting of rhinitis, otitis media, laryngitis, bronchitis, and tracheitis. Advanced disease is characterized by bronchiectasis, bronchiolectasis, pulmonary abscesses, and pneumonia. Animals carrying *M. pulmonis* are not suitable for research because they may be clinically ill and have altered immune responses, predisposing them to other infections. *M. pulmonis* can complicate diagnosis of hemolymphatic neoplasms such as pulmonary lymphoma of rats used in lifespan bioassays. This is because of MRM's hallmark characteristic of inflammation (e.g., accumulation of large numbers of lymphocytes and plasma cells), cellular components that are highly mitogenic for rat lymphocytes, and pulmonary lesions profoundly influenced by chemical administration and concurrent infections, among other factors. *Mycoplasma* spp. is a common (animal and human) tumor and cell-line contaminant; however, *M. pulmonis* is rarely isolated.

- Clinical signs: Animals are often asymptomatic but may show nonspecific signs such as sniffling, respiratory distress, torticollis, and infertility. Signs and lesions depend on various factors, including the *M. pulmonis* strain, rat strain, presence of concurrent infections, and environmental conditions. Cage ammonia levels more than 25 ppm and concurrent infections with other pathogens such as Sendai virus and cilia-associated respiratory (CAR) bacillus can exacerbate clinical signs and lesions.
- Transmission: Primarily aerosol; however, direct contact and transplacental transmission occur.

- Gross lesions: There is serous to purulent exudate in the nasal passages, trachea, and occasionally the tympanic bullae. Gray nodules are present in the lungs, particularly in the apical lobes in a "cobblestone" pattern (saccular bronchiectasis, **Figures 33** and **34**). The diaphragmatic lobes may show compensatory emphysema. If the uterus is severely affected,

(a)

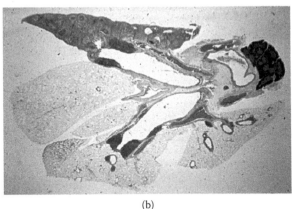

(b)

Fig. 33 Chronic respiratory disease caused by *Mycoplasma pulmonis*. (a) Bronchioles and partial lung parenchymal replacement with pus-filled nodules indicative of chronic suppurative bronchopneumonia and saccular bronchiolectasis. (b) Histopathology demonstrating lymphocytic aggregates (cuffing pneumonia), neutrophils, and necrotic debris. (Images courtesy Department of Comparative Medicine, College of Medicine, Penn State Hershey Medical Center.)

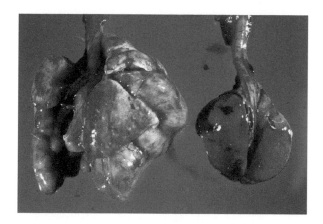

Fig. 34 Chronic respiratory disease. Left: lungs, chronic suppurative bronchopneumonia and saccular bronchiectasis caused by *Mycoplasma pulmonis*. Right: lungs, normal. (Image courtesy Department of Comparative Medicine, College of Medicine, Penn State Hershey Medical Center.)

purulent exudate may be present in the lumen. Lewis rats are highly susceptible to severe genital disease, the primary lesions being purulent endometritis, pyometritis, salpingitis, and perio-ophoritis. Less susceptible strains (e.g., F344), may exhibit no gross genital tract lesions, with decreased reproduction.

- Histopathology: Abundant airway neutrophilia, mucosal epithelium hyperplasia, and lymphoid hyperplasia are present. The hyperplasia of the bronchus-associated lymphoid tissue (BALT) is characteristic of MRM, and has been related to the highly mitogenic cellular components as described above. Squamous metaplasia of the respiratory mucosa may also be present but should be distinguished from that caused by Sendai virus, sialodacryoadenitis virus (SDAV), and hypovitaminosis A. In rats, humoral antibody provides little protection to MRM.

- Diagnosis: Definitive diagnosis is via serology as the organisms persist despite the presence of antibodies. However, animals may be infected for some time, perhaps months, before antibodies develop against *M. pulmonis*. Thus, culture and polymerase chain reaction (PCR) typically on lavage fluid remain useful in detecting early infections. The most common serologic assay used for detecting MRM is the multiplex fluorescent immunoassay (MFI), which is able to detect IgG antibodies for most rodent pathogens. Enzyme-linked immunosorbent assay (ELISA), indirect fluorescent antibody (IFA) assays, and western blot can also be used. Culturing *M. pulmonis* is feasible but requires prolonged incubation, multisite culture, and specialized broths and agars (e.g., Hayflick's media). CAR bacillus is a frequent co-pathogen with *M. pulmonis*, and diagnostic investigations should include screening for this organism, too.

cilia-associated respiratory bacillus (car bacillus)

This unclassified Gram-negative, motile, slightly fusiform, non-spore-forming bacterium has been reported as an etiologic agent of CRD. The organism colonizes the ciliated epithelium of airways and survives freezing and thawing. Signs and lesions are similar to infection with *M. pulmonis*, but co-infection with *M. pulmonis* is also common. Infection with CAR bacillus can easily spread from infected rats to uninfected cagemates within 2 months.

- Clinical signs: Rats may be asymptomatic or show varying degrees of nonspecific signs, including respiratory distress.

- Transmission: Primarily via direct contact; thus, soiled-bedding sentinel rats may fail to detect the organism in an infected colony. Transmission studies indicate that intranasal inoculation of the organism results in infection. Transmission from infected dams to pups has also been reported.

- Gross lesions: Lesions in the lung vary from a mild, patchy mottling to scattered gray raised foci resembling those seen in MRM.

- Histopathology: The predominant lesion is peribronchiolar cuffing with lymphocytes and plasma cells. The bronchial epithelium is normal to slightly hyperplastic and hypertrophic. Bronchiectasis and bronchial abscessation are evident in severe cases.

- Diagnosis: Culture can be performed but is not generally recommended for routine diagnostic purposes. The organism has been grown in embryonated chicken eggs, cell culture, and cell-free media. Routine screening is primarily through serological assays such as MFI, ELISA, and IFA. Silver stains such as Warthin-Starry or Steiner's stain on histopathology and tracheal scrapings will reveal slender, argyrophilic bacilli along the apices of the ciliated respiratory epithelium. PCR on nasal/oral swabs and affected tissues is a confirmatory test.

Corynebacterium kutscheri

This Gram-positive diphtheroid bacillus causes pseudotuberculosis. Rats can harbor *C. kutscheri* in their oropharynx, submaxillary lymph nodes, and large intestine without ill effects or apparent circulating antibodies. Stress, immunosuppression, and nutritional deficiencies may result in disease expression and mortality.

- Clinical signs: The respiratory tract, middle ears, and preputial glands are common infection sites, with nonspecific clinical signs of active disease including respiratory distress and chromodacryorrhea.

- Transmission: Fecal-oral. Experimentally, animals fail to clear the infection and eventually shed the bacteria in the feces for up to 5 mo. Hematogenous spread with dissemination to thoracic and abdominal viscera may also occur.

- Gross lesions: Lesions (**Figure 35**) are primarily pulmonary abscesses with diameters of 0.25 mm to 1 cm. Abscesses may also occur in the liver, kidney, spleen, subcutaneous tissue, and peritoneal cavity.

- Histopathology: Affected organs exhibit multifocal coagulative or caseous necrosis with neutrophilic infiltrates. Mononuclear inflammatory cells are found in older lesions. There may be pulmonary edema and perivascular cuffing. Organisms are readily apparent in suppurative lesions.

- Diagnosis: Samples of affected organs (submandibular cervical lymph node is the best site) or tracheal washings should be taken for bacterial culture. To detect carriers and subclinical cases, oropharyngeal swabs can be performed but may be ineffective. FCN medium (a brain-heart infusion medium with furazolidone, nalidixic acid, and colimycin to inhibit the growth of Gram-negative rods) is the medium of choice for *C. kutscheri* culture. A preliminary diagnosis can be made on finding typical Gram-positive rods with "Chinese letter" configurations on impression smears taken at necropsy. Warthin-Starry (silver) or Giemsa stains can also be used.

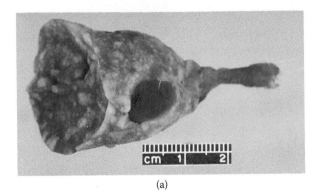

(a)

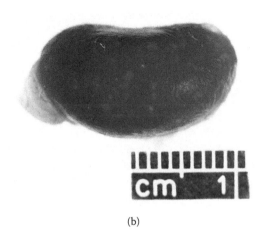

(b)

Fig. 35 Typical *Corynebacterium kutscheri* lesions found in the rat lungs (a), and the rat kidney (b). The lungs contain multifocal pale areas of varying diameter.

There is no commercially available serology for *C. kutscheri*, although an ELISA test has been developed. PCR is another possible diagnostic method, but is not widely commercially available.

Pasteurella pneumotropica

This Gram-negative bipolar-staining coccobacillus is an opportunistic pathogen readily colonizing the intestine and may be found in the conjunctiva, nasopharynx, lower respiratory tract, and uterus. Although prevalent in many rodent colonies, it is of low virulence. Most infections are clinically inapparent, and very few research interactions have been documented in asymptomatic infections. There are few reports of *P. pneumotropica* as a primary pathogen. In the presence of primary pathogens, *P. pneumotropica* potentiates the severity of disease. It is a co-pathogen resulting in pneumonia and otitis media with *M. pulmonis* and in embryonic death with the Sendai virus.

- Pathology/Clinical signs: Pathology is indistinct, and the organism is recoverable in grossly normal appearing tissues. There may be ocular or nasal discharge, conjunctivitis, otitis media, head tilt, dyspnea, skin abscesses and mastitis, purulent exudate within the tympanic bullae and nasal passages, bronchopneumonia, and pyometra.
- Transmission: Direct contact and fecal-oral route, rather than by aerosols. Vertical transmission can occur.
- Diagnosis: The bacteria can be isolated from affected organs and grows well on basic media such as blood agar plates. Nasopharyngeal washes and swabbing techniques are effective in isolating *P. pneumotropica* from young and older animals, respectively.

Streptococcus pneumoniae

This Gram-positive coccus is also referred to as diplococcus or pneumococcus. It is ubiquitous in humans and animals, with rats being asymptomatic carriers for several serotypes, including 2, 3, and 19. Although humans and rats carry some of the same serotypes, *S. pneumoniae* is not considered a true zoonotic agent. It may cause acute primary disease with mortality, but is more important as a secondary invader. The bacteria are carried in the nasal turbinates and the tympanic bullae, and the infection often remains localized. With stress (e.g., shipping, environmental changes, or experimental manipulation) and concurrent infection, overt disease may develop.

- Clinical signs: Serous to mucopurulent nasal discharge is common, as is chromodacryorrhea, head tilt, and dyspnea. There may be mortality. Very young and old rats are most susceptible.
- Gross lesions: In severe cases, lungs are firm and red, with a thick white exudate in the thoracic cavity. The pleura and pericardium are often thickened with fibrin and purulent

exudate. Fibrinopurulent exudate may also occur in the abdomen, genital system, and meninges of the brain.

- Histopathology: Lesions include fibrinopurulent bronchopneumonia, pleuritis, pericarditis, peritonitis, orchitis, or meningitis.
- Diagnosis: The organism is grown on 5% blood agar in a 10% CO_2 atmosphere. Alpha hemolysis (seen as a green tint) is produced around the bacterial colony on blood agar plates. There are many nonpathogenic alpha-hemolytic streptococci, and identification is through use of optochin disks (hydrocuprein hydrochloride) on sample cultures. Optochin inhibits the growth of S. *pneumoniae*, but allows growth of the nonpathogenic streptococci. Serologic typing is based on capsular swelling and is called the *Neufeld Quellung reaction*. Gram staining of smears from lesions will reveal typical encapsulated diplococci.

Bordetella bronchiseptica

This Gram-negative, small rod-shaped bacterium is a primary rabbit and guinea pig pathogen. In rats, it is an opportunistic pathogen causing respiratory infections characterized by suppurative rhinitis and bronchopneumonia.

Clostridium piliforme (Tyzzer's Disease)

Clostridium piliforme is a Gram-negative spore-forming rod causing Tyzzer's Disease, an enterohepatic disease with secondary involvement of the heart. It is an intracellular obligate anaerobe, and so cannot be cultivated on artificial media. It is widely distributed in many rodent species and rabbits but isolates tend to be host-specific. Young adults are most often affected, primarily in conditions of stress, poor husbandry, or immunosuppression.

- Clinical signs: Nonspecific clinical signs include lethargy, weight loss, and distended abdomen, but clinically inapparent infections do occur. Diarrhea and acute death with no clinical signs may also be seen.
- Transmission: Spores are ingested from the environment or feces of an infected animal. Transplacental transmission is possible. The spores are relatively stable in the environment and can remain viable in contaminated bedding for up to 1 year.

This organism is susceptible to disinfectants such as sodium hypochlorite, and spores are inactivated at 80°C for 30 min.

- Gross lesions: The major gross lesions involve the liver, ileum, and myocardium ("Tyzzer's triad"). The hallmark of infection is flaccid segmental terminal small intestinal dilatation (megaloileitis). Ileal lesions may extend to adjacent cecum and jejunum. In the liver, there are few to many scattered pale foci up to several millimeters in diameter. Circumscribed grayish foci may be seen in the myocardium in some cases. The mesenteric lymph nodes are usually swollen.

- Histopathology: Hepatic lesions include focal hepatitis and/or foci of coagulative necrosis with a distinct boundary between necrotic and adjacent normal tissue. The organism is found in viable hepatocytes on the periphery of the necrotic foci. Myocardial lesions vary in size, ranging from necrosis of a few fibers to complete transmural involvement. In the intestine, there is frequently a necrotizing transmural ileitis. Inflammatory infiltrates are variable.

- Diagnosis: Definitive diagnosis is based on finding the organism and lesions in affected tissues ("Tyzzer's triad"). Special stains such as Gram stain, Giemsa, and methylene blue are used for impression smears. For tissue stains, the Warthin-Starry, Giemsa, or PAS are valuable. Serological assays (MFI, ELISA, and IFA) are available as screening tools and have good negative predictive value. However, owing to the complex nature of bacterial antigens, false positive results are possible. If serology is positive, a confirmatory stress test by injecting rats with cyclophosphamide and exposing them to *C. piliforme* spores can be performed. If test rats do not present with clinical disease, then the serology was a false positive, or the animals may have a nontoxigenic strain of *C. piliforme*. *C. piliforme* grows in embryonated hen's eggs and some mammalian cell cultures, but most laboratories are unable to directly culture it. Fecal PCR is available and can be used to determine active shedding; however, the organism is cleared from immunocompetent animals. PCR of lesions is also a useful adjunct to histopathology. One must distinguish the ileal distention of Tyzzer's disease from adynamic ileus associated with intraperitoneal injection of the anesthetic chloral hydrate.

Enterococcus durans (Streptococcal Enteropathy of Infant Rats)

This organism causes enteropathy in infant (suckling) rats resulting in high morbidity and mortality.

- Clinical signs: Diarrhea and stunted growth.
- Gross lesions: Abdominal distention and fecal soiling. Stomachs are distended with milk, with concurrent dilation of the small intestine.
- Histopathology: There is little change in the intestinal villi and minimal or no inflammatory response. However, large numbers of Gram-positive cocci are typically present over the villar surface.
- Diagnosis: Culture and isolation.

Staphylococcus aureus

S. aureus is the most frequently isolated bacterial species from rats. Coagulase-positive *S. aureus* has been associated with ulcerative dermatitis with hair loss in rats. Lesions occur over the shoulders and rib cage, submandibular region, neck, ears, and head. Histopathology suggests an epidermolytic toxin is an important factor in the condition, as early stages are suggestive of burn lesions. The lesions are most likely due to self-trauma and inoculation of the wound with bacteria.

Pseudomonas aeruginosa

P. aeruginosa is a Gram-negative opportunistic bacterial rod colonizing the rat oropharynx. The organism is most commonly recovered from water conduits and human carriers. It is primarily a nosocomial infection and has been a major problem in neutropenia-inducing research procedures, including irradiation, burns, and treatment with immunosuppressants such as steroids. Surgical procedures such as implantation of indwelling jugular catheters are subject to developing pseudomoniasis. Antibiotic treatment may facilitate *P. aeruginosa* colonization. In immunocompromised animals, there may be facial edema, conjunctivitis, and nasal discharge. Contaminated water bottles and automatic watering lixits are common bacteria sources. The organism grows well on blood agar, and may produce a blue-green pigment (pyocyanin or fluorescein).

Helicobacter spp.

Helicobacter has emerged as a common laboratory rodent pathogen, with prevalence data from diagnostic laboratories approaching 20% for some *Helicobacter* species. These Gram-negative rods are spiral, fusiform, or curved, and flagellated. *H. bilis*, *H. muridarum*, *H. rodentium*, *H. trogontum*, *H. hepaticus*, *H. ganmani*, and *H. typhlonius* have been isolated in rats; with *H. bilis* being associated with clinical disease. *Helicobacter* primarily colonizes the cecum and colon, although some species colonize the gall bladder, liver, or stomach. The typhlocolitis caused by enterohepatic *Helicobacter* species resembles inflammatory bowel disease.

- Pathology/Clinical signs: Most animals carrying *Helicobacter* spp. are asymptomatic, but immunodeficient animals are susceptible to disease. For example, *H. bilis* is associated with proliferative and ulcerative typhlitis, colitis, and proctitis in athymic nude rats. Currently, there is no convincing evidence that *Helicobacter* causes clinical disease in immunocompetent rats.

- Transmission: Fecal-oral. Although vertical transmission has not been reported, *H. typhlonius* DNA has been detected in the sex organs of some mouse strains. Transmission by tumor transplantation has also been reported. The organism is highly sensitive to desiccation and does not transfer readily from one room to another.

- Diagnosis: Although serologic testing for *Helicobacter*-specific serum IgG or mucosal IgA antibody responses using ELISA and western blot are reported, there are currently no commercially available *Helicobacter* serology assays principally because of the lack of a universal antigen capable of detecting all *Helicobacter* species. Histology using silver stains such as Warthin-Starry is available to visualize the bacteria but has low sensitivity. More importantly, several PCR assays including basic fecal and fluorogenic nuclease PCR assays have been standardized. Detection by PCR is currently the "gold standard" owing to very high sensitivity coupled with the option of using alternate primers to obtain species-level identification.

- Treatment/Control: Special diets for oral dosing of drug combinations for eradication of *Helicobacter* are commercially available. The feed has either a combination of omeprazole, metronidazole, amoxicillin, and clarithromycin, or one of bismuth, metronidazole, and amoxicillin. Once the antibiotic is withdrawn, there is a high reoccurrence. Various rederivation techniques are effective, with cross fostering up to four days following birth being effective in mice.

viral agents

Rat Parvoviruses

These single-stranded, nonenveloped DNA viruses are relatively common in laboratory and wild rats primarily owing to the viruses' ability to persist in infected animals and the environment, including resistance to nonoxidizing disinfectants. There are three major genetic/antigenic groups of parvoviruses in rats. The first two groups are related; their prototypes are Kilham's rat virus (RV) and Toolan's H-1 virus (H-1). The third group is called rat parvovirus (RPV), previously called rat orphan parvovirus (ROPV). If infected with RV as infants, rats may be persistently infected with a high antibody titer up to 14 weeks. Large doses and experimental infections of RV are models for teratologic effects, cerebellar hypoplasia, hepatitis, and hemorrhagic encephalopathy. There are no reports of naturally occurring disease associated with H-1 viruses or RPVs. Parvoviruses have a tropism for rapidly dividing cells, especially lymphoid tissue. Research effects of parvovirus infection in rodents include long-lasting effects on the immune system, oncology studies, and lymphocyte cultures. For example, infection can lead to autoimmune diabetes in diabetes-resistant BB rats.

- Clinical signs/Gross lesions: Clinical signs are frequently subclinical in immunocompetent animals, but RV infection may produce disease with natural infections in naïve and immunosuppressed rats. RV findings vary with age. Pups are runted and may have ataxia, cerebellar hypoplasia, and jaundice. In adults, there may be jaundice, and hemorrhage and cyanosis within the scrotum, poor body condition with loss of body fat, and congestion of lymph nodes. Pregnant rats may have an increase in the number of resorption sites. Young adults may be dehydrated with abdominal swelling.

- Histopathology: RV infection causes hemorrhagic infarction with thrombosis in multiple organs, including the brain, spinal cord, testes, and epididymis; and multifocal hepatic necrosis. Rat pathology is not enteric as they lack enterocyte parvoviral receptors.

- Transmission: Primarily fecal-oral through the horizontal route, direct contact, or fomites (including contaminated feed and bedding). Animals shed virus in urine, feces, milk, and oronasal secretions. RV may be transmitted transplacentally, resulting in poor reproductive parameters.

- Diagnosis: Serologic testing, including MFI, virus neutralization (NT), ELISA, and IFA, are routinely used as screening tests. Specific assays are available for structural antigens (VP) specific to each parvovirus and nonstructural (NS) antigens common to all parvoviruses. Diagnosis can be confirmed by PCR on tissue or feces, and demonstrating tissue antigen using immunocytochemistry. Mesenteric lymph nodes or spleen are the preferred tissues for testing, but RV is also readily detected in the lung. PCR or the rat antibody production (RAP) test can be used on murine biological products. Hemagglutination inhibition (HI) assay is not suitable for RPV diagnosis.

Note: Contaminated cell lines and tumors are important sources of parvoviral infections, primarily because of parvoviruses' prevalence, environmental stability, and requirement for dividing cells for replication.

Sialodacryoadenitis Virus (SDAV)/Rat Coronavirus

Coronaviruses are enveloped and include rat coronavirus, or Parker's rat coronavirus (PRC), and SDAV. PRC and SDAV properties are very similar, with both viruses replicating in the respiratory tract and producing infections exhibiting high morbidity, low mortality, and a rapid spread. Coronoviral infections are now relatively uncommon in modern laboratory animal facilities, but common in pet rats. PRC and SDAV share common antigens and cross-react with some mouse hepatitis virus strains. SDAV's tropism for tubuloalveolar glandular tissue of serous or mucoserous glands severely impacts the serous salivary and lacrimal gland product.

- Clinical signs: SDAV is highly infectious, but uncomplicated infections rarely result in death. Most rats show clinical signs a few days following exposure. Signs usually regress within one week and lesions regress in 2–4 weeks. Clinical signs are sneezing, photophobia, blepharospasm, epiphora, swelling of the neck region (impacting the submandibular salivary glands), and chromodacryorrhea. Some rats develop chronic keratoconjunctivitis, with corneal opacity, ulcers, pannus, synechia, hypopyon and hyphema, and cataracts **(Figure 36)**. Ocular changes result from the destructive lesions and impaired lacrimal gland function. Neonatal mortality and estrous cycle aberrations are associated with SDAV infection. Infection in immunodeficient rats can persist with severe clinical signs and may be fatal.

- Gross lesions: Rhinitis, inflamed and edematous lacrimal and salivary glands, and enlarged lymph nodes.

- Histopathology: Necrosis and inflammation associated with the infection are self-limiting and repairable. The nasal turbinates are most severely affected, with the olfactory epithelium usually spared. There are mild, similar tracheal lesions. In the lungs, changes include mild hyperplasia of the peribronchiolar lymphoid nodules. Salivary gland lesions are first characterized by necrosis and inflammation followed by squamous metaplasia during the repair phase. Oropharyngeal serous and mixed salivary glands are affected, with mucousproducing salivary glands being resistant. Repair begins during the second week and is usually complete by the third week. In athymic (nude) rats, infections and lesions may persist up to 6 months, characterized by chronic suppurative rhinitis and bronchopneumonia, with chronic salivary and Harderian gland inflammatory lesions.

- Transmission: Aerosol or direct contact with infected nasal or salivary secretions. The virus fails to persist in immunocompetent rats.

- Diagnosis: Identification of typical clinical signs and histopathology during the first week of infection are sufficient to make the diagnosis. MFI, ELISA, and IFA are common serologic diagnostic tests. Immunocytochemistry with demonstration of the antigen and PCR for salivary or lacrimal tissue of acutely infected rats can be performed. The virus is undetected in the cornea or other ocular components. SDAV infection should be differentiated from *Mycoplasma*, Sendai virus, pneumonia virus of mice, periorbital bleeding sequelae, and irritation from high ammonia levels or stress-associated chromodacryorrhea.

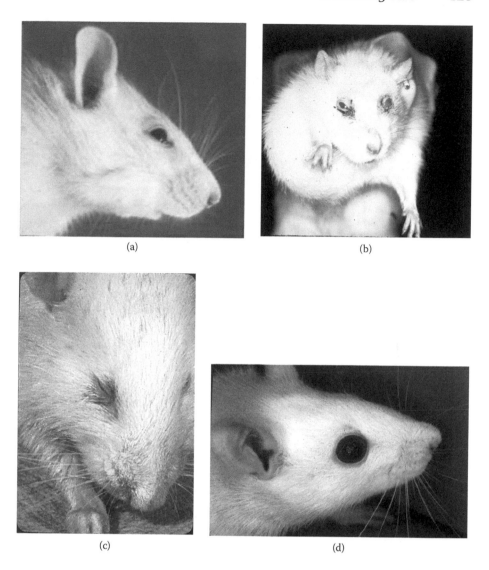

(a)

(b)

(c)

(d)

Fig. 36 Sialodacryoadenitis virus. SDAV clinical manifestations include (a) ventral cervical swelling; (b) chromodacryorrhea and ulcerative hemorrhagic keratitis; (c) blepharospasm and chromodacryorrhea; and (d) ulcerative and hemorrhagic keratitis. (Images courtesy Department of Comparative Medicine, College of Medicine, Penn State Hershey Medical Center.)

PRC primarily affects the lungs, usually with no gross lesions. The microscopic lesion is characterized by patchy interstitial pneumonia, which is mild and short-lived. There may be transient rhinotracheitis. Salivary gland lesions are uncommon and, when present, are mild and similar to SDAV. The lesions apparently do not impact the lacrimal gland.

Pneumonia Virus of Mice (PVM)

This enveloped RNA virus of the genus *Pneumovirus*, family Paramyxoviridae, typically causes an asymptomatic infection with some rats developing a transient, mild interstitial pneumonia. It is not as infectious as other paramyxoviruses but is transmitted by aerosol and direct contact with respiratory secretions. There are usually no clinical signs or gross lesions, but there may be focal to multifocal plum-colored to gray foci, less than 2 mm diameter in any lung lobe. In acute stages, typical microscopic lesions include nonsuppurative vasculitis and interstitial alveolitis with necrosis. Differential diagnoses include Sendai and rat coronavirus. Diagnosis is by histology and serologic assay, primarily MFI, ELISA, and IFA. Recovery of the virus from the respiratory tract, immunohistochemistry, and PCR of lung tissues is also available.

Sendai Virus (SV)

SV is another *Paramyxovirus*, but from the genus *Respirovirus*. SV is extremely contagious and causes acute respiratory infection, which "dies out" in 4 to 8 weeks in immunocompetent rats if no new animals are introduced into the colony. The virus replicates in type I and II pneumocytes and in alveolar macrophages. Infection can be superimposed and contribute to respiratory lesions caused by *M. pulmonis*, and presumably *P. pneumotropica*. It can mimic changes seen with exposure to halogenated aromatic hydrocarbons, oxidant gases, and other toxicants. It may impair normal immune responses and fetal development. There are considerable differences between rat strains concerning SV susceptibility, with LEW and BN rats being more susceptible than F344. Transmission to mice, guinea pigs, and hamsters is possible.

- Clinical signs: Because of animal strain susceptibility differences, pathologic lesions vary in severity; however, natural rat infections are generally asymptomatic, with only minor effects on reproduction and pup growth. Nonspecific clinical signs including rough coat, dyspnea, and anorexia can be seen.

- Gross lesions: Usually few, although patchy pulmonary consolidation may be seen.
- Histopathology: There is focal nonsuppurative interstitial pneumonia, with mild to severe peribronchiolar and perivascular cuffing, persisting for several weeks. Rhinitis with focal diffuse necrosis of the respiratory epithelium also occurs. Necrotizing bronchitis is reported in experimentally infected germ-free rats.
- Transmission: Aerosols or direct contact; the virus is not transmitted well by soiled bedding.
- Diagnosis: Confirmed by histologic lesions and serologic titers (MFI, ELISA, and IFA), although antibody levels may drop below detectable range after 9 months. PCR is recommended on symptomatic animals, as these animals may not have seroconverted to SV.

Rotavirus

This unenveloped RNA virus of the family Reoviridae (group B rotavirus) causes the nonlethal condition "infectious diarrhea of infant rats" (IDIR).

- Clinical signs: Poor growth, diarrhea, and perianal dermatitis in suckling rats.
- Gross lesions: The stomach and the proximal small intestine usually contains milk curd and watery contents, respectively. The distal small intestine and large intestine contain yellow-brown to green fluid and gas.
- Histopathology: Enterocyte vacuolation and necrosis, villi blunting, and eosinophilic intracytoplasmic inclusions in epithelial syncytia, especially in the ileum of rats less than 14 days of age.
- Transmission: Primarily via the fecal-oral route, with fomites, environmental particles, and possibly human contact playing a role.
- Diagnosis: Serology and fecal PCR may be used to investigate suspected outbreaks.

Rat Theilovirus (RTV)

RTV is a nonenveloped RNA virus in the family Picornaviridae, genus *Cardiovirus*, and species *theilovirus*. It is closely related to, but distinct from, Theiler's murine encephalomyelitis virus (TMEV) strains. Fecal-oral transmission occurs. Little is currently known about RTV. A study revealed SD rats were more susceptible than CD rats to RTV infection based on the fecal shedding period and antibody production, suggesting SD rats are the more effective RTV sentinels. No clinical signs or lesions are observed in rats infected with RTV, although a single report found neonatal rats inoculated with intestinal homogenates from rats with antibodies to the TMEV strain GDVII developed neurologic signs: flaccid hindlimb paralysis and tremors. Regarding research complications, RTV should be considered similar to TMEV until further data are compiled on its effects in rats. TMEV can potentially interfere with research involving the nervous, immune, and musculoskeletal systems. RTV infection is diagnosed through serology (ELISA, IFA, MFI). PCR is also available.

Other Viral Infections

There are other infrequently occurring wild and/or laboratory rat viral infections, including DNA viruses such as rat cytomegalovirus and polyoma virus. Rats seroconvert to mouse adenovirus (MAD) 1 and 2 and reovirus 3 (REO3), but disease from natural and experimental infections is unreported. REO3 and MAD 1 and 2 can be included in routine colony health serology surveillance testing.

parasitic agents

Ectoparasites

Reports of mite and lice infestations in laboratory rodent facilities have generally been uncommon in the last two decades. However, mite infestations have recently been reported in several institutions and in wild urban rat populations. Rat fleas (e.g., *Xenopsylla*) are rare in laboratory rats.

Mites

Ornithonyssus bacoti, the tropical rat mite, and *Laelaps echidninus*, the spiny rat mite, are rarely encountered in laboratory animal setting. Both mites remain on the host only long enough to obtain a blood meal. They are not host-specific, will bite humans, and can act as vectors for several agents pathogenic to humans. *O. bacoti* is the most commonly reported mesostigmatid mite both in laboratory rodent colonies and in human-associated rat mite dermatitis cases. The mites are on the host only when feeding, after which they seek refuge in the surrounding environment. They can cause anemia, debility, and infertility in rats.

Radfordia ensifera, the rat fur mite, has a limited host range and feeds on skin debris. Clinical signs may not be observed; but in heavy infestations, alopecia and self-induced trauma from scratching are common. Transmission is by direct contact.

Lice

Rats are host to two louse species, *Polyplax spinulosa* (spined rat louse) (**Figure 37**) and *Hoplopleura pacifica* (tropical rat louse), of which only the former has been widely described. *P. spinulosa* is a blood-sucking louse completing its life cycle on the host. It causes unthrifty appearance, pruritus, small skin wounds, and anemia directly by feeding and indirectly by acting as a *Hemobartonella muris* (*Mycoplasma haemomuris*) vector. Transmission is by direct contact.

Fleas

Several flea genera, including *Xenopsylla*, *Leptopsylla*, and *Nosopsyllus*, infest wild rats and, rarely, laboratory rats. Laboratory rat transmission occurs when wild rats access laboratory facilities.

Diagnosis

O. bacoti and *L. echidninus* require microscopic pelt and bedding examination for the mites. *R. ensifera* and *P. spinulosa* diagnosis

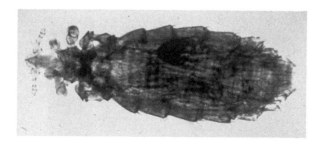

Fig. 37 *Polyplax* ssp. female. This louse can cause anemia and generalized skin lesions; it is yellow-brown in color and may be found around the animal's shoulder and neck region and measures from 0.7 to 1.5 mm in length. (Image courtesy Gail Moore and Suellen Greco.)

is by skin scraping or plucking hairs and examining the specimen microscopically for the mites. If the animal is examined postmortem, the pelt can be removed and placed in a Petri dish or on black paper. As the pelt cools, the parasites become visible through a dissecting microscope as they migrate to the ends of the hair. Some mites may also be evident in tissue sections of affected skin. For fleas, diagnosis is made by identifying the parasite on the host.

Endoparasites

Nematodes

Syphacia obvelata, *S. muris*, and *Aspiculuris tetraptera* are round-worms (pinworms) inhabiting the rat colon and cecum. They have a direct life cycle (*Syphacia*: 11–15 days; *Aspiculuris*: 23–25 days). Adult *Aspiculuris* are readily recognized by the four alae present at the anterior end of the body. Eggs are bilaterally symmetric, 89–93 × 36–42 μm in size, and are passed in the feces. *Syphacia* adults of both species look similar, but *S. muris*, the rat pinworm, is slightly smaller and the male has a longer tail, measured as a proportion of body width. Identification of *Syphacia* is primarily by eggs (**Figure 38**) deposited in the perianal region. *S. obvelata* eggs are almost completely flat on one side and measure 118–153 × 33–55 μm, while *S. muris* eggs are only slightly flattened on one side and measure 72–82 × 25–36 μm. *S. muris* infection is usually asymptomatic; but soft stool, enteritis, perianal irritation, impactions, rectal prolapse, and intussusceptions are possible. Microscopically there is

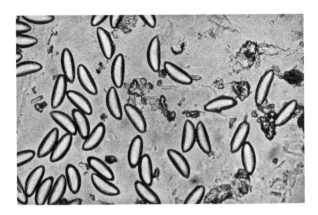

Fig. 38 *Syphacia muris* eggs, the rat pinworm, are banana-shaped and easily seen on fecal floatation or cellophane tape test (shown). Eggs measure 72–82 μm (length) × 25–36 μm (width).

little pathology, although a multifocal granulomatous reaction in the lamina propria may be observed in the large intestine.

Diagnosis: *Syphacia* may be diagnosed by cellophane tape impression, made by touching clear cellophane tape to the perianal region and placing the tape on a glass slide for microscopic examination. *Aspiculuris* (**Figure 39** to **Figure 41**) is best diagnosed by fecal flotation. Cecal contents may be directly examined for adult worms. Fecal PCR is available, and is highly specific and as sensitive as perianal tape test and fecal flotation. Adult worm cross sections may be identified in histologic sections of the cecum or colon. The two *Syphacia* may be speciated by the location of three rounded structures called *mamelons* found in males; the anterior mamelon in *S. muris* is located near the center of the body, whereas in *S. obvelata* the middle mamelon is near the center of the body (**Figure 42**).

Cestodes

Tapeworms are common in wild rodents but rare in laboratory populations.

Rodentolepis nana (*Hymenolepis nana*), the dwarf tapeworm, may be found in rodent small intestines and is of zoonotic concern because it can be transmitted directly, without employing an intermediate host as other cestodes do. Clinical signs depend on the parasitic burden, with heavy infections causing severe catarrhal enteritis.

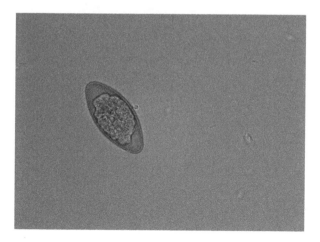

Fig. 39 High power magnification of *Aspiculuris tetraptera* ovum demonstrating the typical egg morphology. Ova measure 89–93 μm in length × 36–42 μm in width. (Image courtesy Scott Trasti and Anna Acuna.)

Fig. 40 Low power magnification of an *Aspiculuris tetraptera* female. Males typically measure 2–4 mm long × 120–190 µm wide, while females are 3–4 mm long × 215–275 µm wide.

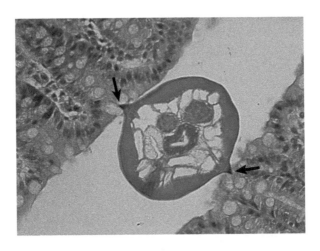

Fig. 41 High power histopathology section of *Aspiculuris tetraptera* in cross section. Note the lateral alae (black arrows). (Image courtesy Scott Trasti and Anna Acuna.)

Physiological effects include a decline in rat serum albumin levels for the first 20 days postinfection.

Hymenolepis diminuta, the rat tapeworm, inhabits the upper small intestine of rodents and primates. Rat strains differ in susceptibility to infection, with TM and DA strains developing 60% and 30% fewer total adult worms, respectively, than other strains. Light infections

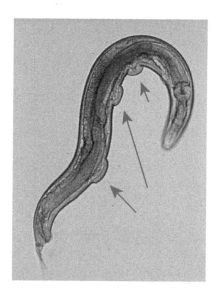

Fig. 42 Male *Syphacia obvelata* (mouse pinworm) demonstrating the three mamelons (arrows). Note the middle mamelon (long arrow) is found near the longitudinal center of the body. *Syphacia muris* males measure 1200–1300 μm × 100 μm. (Image courtesy Dawn Shaffer and Craig Franklin, IDEXX-RADIL™.)

with *H. diminuta* are nonpathogenic, while heavy infections may cause acute catarrhal enteritis or chronic enterocolitis with lymphoid hyperplasia.

Rats serve as intermediate hosts of *Taenia taeniaeformis*, the cat tapeworm. The cysts are found in the liver of infected rats. In one report, *Cysticercus fasciolaris*, the larval stage of *T. taeniaeformis*, induced fibrosarcomas that invaded the serosa of multiple organs and extended through the diaphragm into the pleural cavity. Sarcoma development is attributable to chronic inflammatory reaction to the parasite's capsule. Food and bedding contamination by cat feces is a common mode of transmission.

Protozoans

There are many protozoan species known to infect rats, and most are considered nonpathogenic under normal conditions. In a recent survey, *Hexamastix*, *Entamoeba*, trichomonads, *Chilomastix*, and *Spironucleus* are the more prevalent rat protozoa. Diagnosis is by microscopic examination of fecal wet-mount preparations, cecal scrapings, or intestinal histologic sections.

fungal agents

Pneumocystosis

Pneumocystis carinii is a ubiquitous opportunistic organism found in many mammalian species, including rats and mice. Historically considered a protozoon, it has been reclassified as a fungus.

- Clinical signs: Although *P. carinii* is normally not pathogenic, it can cause pneumonia in immunosuppressed or immunocompromised hosts. Recent studies revealed *P. carinii* as the cause of the idiopathic lung lesions in immunocompetent rats previously attributed to rat respiratory virus (RRV), a virus once classified as a member of the *Hantavirus* genus. Clinical signs of pneumocystosis include wasting, rough coat, dyspnea, cyanosis, and death.

- Gross lesions: Focal to diffuse pulmonary consolidation, failure to collapse, and an opaque pale pink color of the lungs.

- Histopathology: Alveolar septal thickening and alveolar filling with foamy, eosinophilic material, presenting a honeycomb appearance. Dense perivascular lymphocytic cuffs and a lymphohistiocytic interstitial pneumonia may be present.

- Transmission: Inhalation of infective cysts.

- Diagnosis: Definitive diagnosis is by PCR of lung tissue. Serology is available as a routine screening test, and PCR on oral swab and bronchoalveolar lavage specimens can be performed for an antemortem diagnosis. Characteristic pulmonary changes in histopathology and special stains using Gomori's methenamine silver (GMS), toluidine blue, periodic acid-Schiff, or silver stain can detect *P. carinii* trophozoites and yeast-like cysts, providing a presumptive diagnosis. Immunohistochemistry is available.

Another *Pneumocystis* species infecting rats is *P. wakefieldiae*, formerly known as *Pneumocystis carinii* f. sp. *ratti*. It is morphologically similar to *P. carinii*, and both frequently co-exist within the same alveoli. Lower environmental temperatures and an increased rat census are associated with *P. wakefieldiae* alone, while higher relative humidity and higher organism lung burdens were associated with infections from *P. carinii* alone.

Encephalitozoon cuniculi

This obligate intracellular microsporidian parasite is found in the brain and kidneys of rats, mice, and rabbits. Infection is often asymptomatic and fails to cause clinical disease in rats. However, it can induce focal granulomatous encephalitis and nephritis, particularly in rabbits. Clusters form in the brain, kidneys, heart, muscles, pancreas, liver, spleen, and other organs. The life cycle is direct, with transmission by ingestion of the environmentally resistant spores. Several sporadic laboratory rat infections are described, and at least three strains of the organism have been identified. A rat isolate has also been successfully propagated *in vitro*. Diagnosis relies on serology (MFI and ELISA) and histopathology, with spore capsules staining positively with Gram, Giemsa, and Goodpasture's carbol fuchsin stains.

Dermatophytosis

In laboratory rats, this is now relatively rare and is usually caused by *Trichophyton mentagrophytes*. Rats may be asymptomatic or demonstrate florid dermal lesions, characterized by patchy hair loss, erythema, and scaling. Diagnosis is by skin scrapings with wet mount preparations on 10% KOH, fungal culture, and histology by PAS or GMS stains.

neoplastic diseases of rats

Mammary Tumors

Mammary tumors (**Figure 43**) are one of the most common rat neoplasms, especially of female SD rats. The vast majority of mammary tumors are benign fibroadenomas. Malignant adenocarcinomas do occur, but are uncommon. These tumors grow slowly, but can become very large. They may ulcerate or, depending on their location, interfere with normal movement or the ability to reach food and water; any of these are criteria for the animal's euthanasia. Tumors may develop wherever there is mammary tissue, i.e., from the axillary area to the inguinal region on either side of the ventral midline and back. Sex, age, genetic, and endocrine factors may play a role in rat mammary tumor incidence. For example, tumor incidence is markedly reduced in ovariectomized rats. While most cases are in female rats, males occasionally have mammary fibroadenomas, too. Incidence

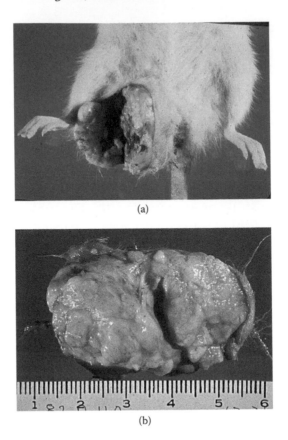

(a)

(b)

Fig. 43 Mammary fibroadenoma. (a) *in situ;* (b) excised. (Image courtesy Department of Comparative Medicine, College of Medicine, Penn State Hershey Medical Center.)

increases with age, especially after 18 months of age. Retroviruses do not appear involved as in mammary tumors of mice. Tumors have a lobulated appearance on cut surface. Surgical removal is feasible, but recurrence in another mammary gland is likely.

Testicular Tumors

The incidence varies greatly between rat strains and colonies. Tumors are discreet, soft, and yellow to brown, with areas of hemorrhage; they may be multiple and are frequently bilateral. The vast majority are benign and include interstitial cell tumors (Leydig cell tumors), especially in F344 and ACI/N rats.

Mononuclear Cell Leukemia, or Large Granular Lymphocyte (LGL) Leukemia

This is a common fatal condition in old F344, Wistar and Wistar-Furth rats. Splenomegaly (**Figure 44**) is a constant finding, with hepatomegaly and lymphadenopathy being variable. Clinical signs include jaundice, anemia, weight loss, and lethargy. The exact cell of origin of the F344 rat LGL leukemia is not fully resolved; natural killer (NK) cell characteristics are present in most if not all cases, but studies on cytotoxic activity and surface antigens suggest these leukemias are of a more heterogeneous lymphocytic cell origin. Interested readers are directed to the review by Thomas et al. (2007).

Pituitary Tumors

Pituitary tumors (**Figure 45**) are well defined, spherical, soft, and vary from single to multiple foci to large masses replacing the whole gland; they measure up to 2 cm in diameter and weigh 350 mg or more. The chromophobe adenoma of the pars distalis is most common. Pituitary tumors occur frequently in older rats, particularly in SD and Wistar strains. Most reports, although inconsistent, indicate a higher incidence in females, especially in virgin females. *Ad libitum* feeding of high caloric diet produces a high incidence of spontaneous tumors compared to a decreased caloric and protein intake. It is not unusual for pituitary and mammary gland tumors to occur in the same animal. Clinical signs vary, but include head tilt, behavioral changes, and sudden death.

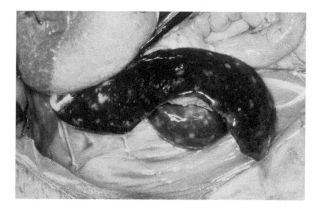

Fig. 44 Splenomegaly with focal raised pale nodules; common in F344 rats with large granular lymphocyte (LGL) leukemia.

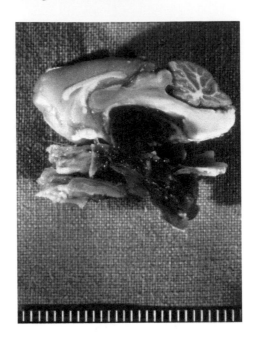

Fig. 45 Pituitary (lactotroph) adenocarcinoma of the pars distalis (bar = 10 mm). (Image courtesy Department of Comparative Medicine, College of Medicine, Penn State Hershey Medical Center.)

Zymbal's Gland Tumor

This tumor affects the holocrine gland located at the base of the external ear (**Figure 46**). Grossly, the tumor is usually a circumscribed mass, frequently with ulceration of the overlying skin. The tumor may be either an adenoma (benign) or adenocarcinoma (malignant). The tumor is relatively rare as a spontaneous tumor but could be induced by certain chemicals.

Other Neoplasms

Keratoacanthoma is a benign skin neoplasm most commonly developing over the chest, back, or tail. The tumor's center is typically filled with a keratin plug. Other neoplasms reported in rats include histiocytic sarcoma, lymphoma, and mesothelioma. Mesothelioma is the most common peritoneal neoplasm in F344 rats, originating from the tunica vaginalis and the ovarian bursa in males and females, respectively.

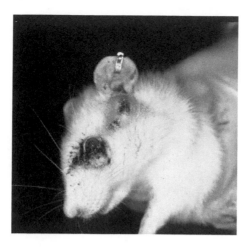

Fig. 46 Zymbal's gland tumor affecting the base of the left ear and chromodacryorrhea. (Image courtesy Department of Comparative Medicine, College of Medicine, Penn State Hershey Medical Center.)

age-related lesions and miscellaneous conditions

Chronic Progressive Nephropathy (CPN)

Also called "old rat nephropathy," CPN is a major old age disease. Although the pathogenesis remains undetermined, several factors have been implicated to influence CPN development, including sex, rat strain, age, diet, and endocrine factors. An earlier onset and more rapid progression occur in male rats. Albino strains and stocks are particularly predisposed, with SD and F344 exhibiting an earlier onset and higher incidence and severity than the Wistar, Brown Norway, and Long-Evans rats. Axenic rats also do not develop significant CPN. Lesions begin as early as 3 months of age but are most extensive in animals at least 12 months of age. Dietary factors, especially *ad libitum* overfeeding and increased dietary protein, significantly impact CPN development. High prolactin levels have also been implicated. CPN is eventually fatal, and is the leading cause of death in old but otherwise healthy rats. The most common clinical signs include renal failure, wasting, and general lethargy. On necropsy, kidneys are pale, irregular, and swollen with a pitted surface **(Figure 47)**. On microscopic examination, there is multifocal to diffuse dilatation of tubules, lined by flattened epithelium. There is interstitial fibrosis and frequently mononuclear cell accumulations.

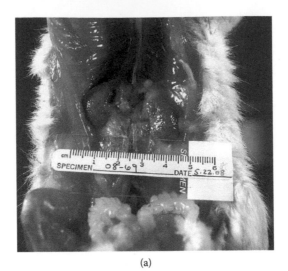

(a)

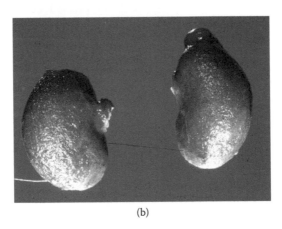

(b)

Fig. 47 Chronic progressive nephropathy. Note the pale, finely pitted kidney surface *in situ* (a) and eviscerated (b). (Image courtesy Dr. Timothy Cooper, Department of Comparative Medicine, College of Medicine, Penn State Hershey Medical Center.)

Glomerular lesions vary from minimal to marked sclerosis. The histological changes result in a number of functional changes, such as increased proteinuria and decreased urine-concentrating ability. CPN is associated with hypertension, polyarteritis nodosa, and renal carcinogenesis, including renal adenomas. For additional information, consult the reviews by Seely and Hard (2008) and Abrass (2000).

Myocardial Degeneration

Common to most stocks and strains (especially SD rats), onset occurs at 12–18 months of age and is more frequent in males. Clinical signs are usually absent. Lesions are usually microscopic, but small grayish foci may be noted grossly. The papillary muscles and their attachment sites in the left ventricular wall are the most frequent degeneration sites. Microscopically, there is degeneration and atrophy with fibrosis and a mononuclear cell infiltrate. Atrial thrombosis is seen as an age-related lesion and may or may not accompany myocardial degeneration. Although myocardial degeneration and fibrosis are common in older rats, there is little to no evidence of cardiac insufficiency. While the pathogenesis is unclear, the lesion could be secondary to chronic renal disease or coronary arteriosclerosis, with myocardial fibrosis possibly being present in their absence.

Urolithiasis

Urinary calculi (**Figure 48**) are incidental findings in older rats. Clinical signs include hematuria, hemorrhagic cystitis, and if obstruction occurs, anuria. Stones are composed of oxalates, phosphate, carbonate, or a mixture. Urinary calculi should be distinguished from mucoid calculi. Mucoid calculi are agonal copulatory plugs excreted into the urethra and bladder. A link between urinary calculi and neoplasia is described in ACI/N and BN/Bi/Rij rats. Laboratory rodent diets were modified to increase the calcium:phosphorus ratio to reduce the renal calculi and calcium-deposit formation seen in female rats. One may attempt surgical removal, but this is seldom warranted in the laboratory rat.

Alveolar Histiocytosis/Alveolar Proteinosis

Pulmonary foam cell aggregations, also known as foamy macrophages or alveolar histiocytes, are common in the lungs of older rats. On necropsy, these aggregates appear as dull pale yellow foci in the subpleural regions. Microscopically, macrophages have abundant cytoplasm, with contents varying from needle-shaped crystals to vacuolated or homogenous, lightly eosinophilic material. There are no associated clinical signs. Impaired mucociliary clearance and excess surfactant production that surpasses breakdown and clearance have been implicated as contributing factors.

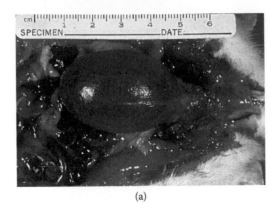

(a)

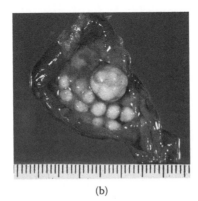

(b)

(c)

Fig. 48 Hemorrhagic cystitis and urolithiasis. (a) The urinary bladder, distended with urine, is hemorrhagic and inflamed from uroliths. (b) Uroliths inside the urinary bladder. (c) Uroliths removed from the urinary bladder. (Images courtesy Department of Comparative Medicine, College of Medicine, Penn State Hershey Medical Center.)

Cholangiofibrosis

This age-related change in the liver is common in F344 and to a lesser extent SD rats. It is uncommon in WAG and BN rats. The cause is unknown. Cystic and telangiectatic lesions also occur in the aged liver. There are usually no clinical signs. Microscopic lesions include portal bile duct proliferation with associated mild fibrosis. A few mononuclear inflammatory cells may be noted. Large inflammatory infiltrates are not generally present.

Radiculoneuropathy

This degenerative condition affects the spinal nerve roots of rats older than 24 months. The cauda equina and ventral spinal nerve roots are most commonly involved, but the degeneration site can vary across strains. The SD, Wistar, BN/Bi/Rij, and WAG/Rij rats are susceptible. There is no treatment. Clinical signs include posterior paresis and paralysis, but some suggest these lesions result from a separate skeletal muscle degenerative process and that the two conditions are unrelated. Microscopically, there is myelin sheath swelling and segmental demyelination. Advanced lesions are associated with axonal degeneration and loss. Although the exact etiology is unknown, neural pressure and hypoactivity may exacerbate this condition.

Polyarteritis Nodosa (PAN)

PAN is a necrotizing vasculitis affecting small- to medium-sized muscular arteries. It is seen in SD, spontaneously hypertensive rats (SHR), and ACI/SegHsd rats, and those rats with late-stage CPN. Clinical signs are inapparent, and when present are nonspecific. However, deaths with hemoabdomen and pancreaticoduodenal artery rupture secondary to PAN are reported. At necropsy, affected vessels, especially the mesenteric vessels, are enlarged and thickened in a segmental pattern with marked tortuosity (**Figure 49**). There is possible vessel involvement of other tissues, including pancreas, kidney, testis, and most other organ tissue except the lung. Affected vessels frequently have aneurysms and thrombi. Microscopically, there is thickening and fibrinoid degeneration of the media of affected arteries. Inflammatory cells are primarily mononuclear, with a few neutrophils. Affected vessel lumens are variable in size and contour and frequently contain thrombi. Although of unknown etiology, the renin-angiotensin system has been associated with PAN development in rats.

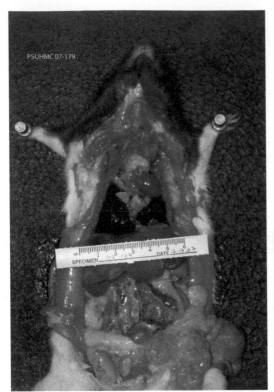

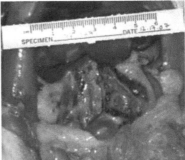

Fig. 49 Polyarteritis nodosa. Thickened, dilated, torturous, hemorrhagic, and inflamed mesenteric arteries. (Image courtesy Department of Comparative Medicine, College of Medicine, Penn State Hershey Medical Center.)

Other Age-Related Diseases

Other rat aging and degenerative disorders include nephrocalcinosis (i.e., renal calcification), hydronephrosis, hematuria and renal papillary hyperplasia, and degenerative osteoarthritis.

other conditions

There are numerous miscellaneous conditions (e.g., hermaphroditism and patent ductus arteriosus) reported in rats. It is beyond the scope of this section to give a detailed description of all these condition. Some can be attributed to or are associated with a rat's particular phenotype/genotype.

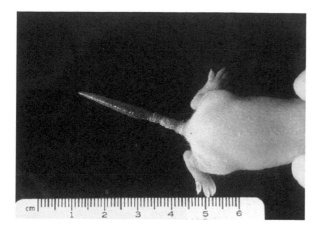

Fig. 50 Ringtail in a pre-weanling rat. Note the distal tail swelling and necrosis and the proximal constrictions. (Image courtesy Department of Comparative Medicine, College of Medicine, Penn State Hershey Medical Center.)

Ringtail

This is characterized by annular tail constrictions, sometimes with hemorrhage and necrosis distal to the constrictions (**Figure 50**). Young rats, especially those raised on wire-bottom caging, are usually affected but may remain clinically normal except for the tail lesions. The condition is classified as an environmental disorder, traditionally attributed to low environmental humidity (< 25%), although dietary deficiencies (e.g., fats, fatty acids, B vitamins, and zinc), genetic susceptibility, environmental temperature, and the degree of animal hydration are suggested causes. Histopathology suggests that epidermal acanthosis and hyperkeratosis are the main and primary events in developing this condition. If tail segments slough, healing usually occurs without further complication. Lanolin ointment has been reported to be effective in treating ringtail in transgenic F344 rats.

Malocclusion

In rodents, malocclusion most commonly refers to overgrown incisors (**Figure 51**). The condition, seen both in young and old rats, results from a variety of hereditary and environmental causes. It is secondary to poor upper- and lower-incisor alignment, which results in abnormal dental wear. It can result from broken teeth, congenital jaw misalignment, and powdered ration feeding. It has been reported as a

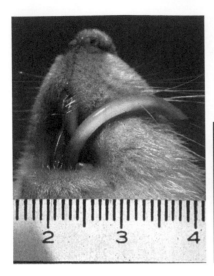 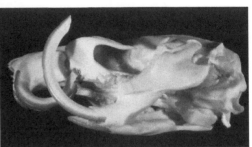

Fig. 51 Malocclusion. Check unthrifty or anorectic animals for malocclusion. (Image courtesy Department of Comparative Medicine, College of Medicine, Penn State Hershey Medical Center.)

spontaneous background lesion in SD and aging Wistar rats. Clinical signs include unthrifty appearance, weight loss, and dehydration. Malocclusion can progress to the point that the teeth may penetrate the opposing gums, causing inflammation and abscessation.

Heat Prostration

This condition most often occurs in the summer months; however, rats placed too close to heating elements (e.g., shipping during the winter months) are equally susceptible. Rats shipped when the ambient temperature is above 85°F (29°C) are prone to heat exhaustion. Clinical signs include increased respiration and excessive salivation, resulting in wet mouths, muzzles, and paws (**Figure 52**). There may be congestion of the nail beds and digits and often a high death loss.

Retinal Degeneration (RD)

Retinal changes are common in albino rats exposed to light intensities ≥325 lux at the cage level. The most severe lesions are seen in rats housed on the top shelves near light fixtures. Lesions vary from decreased photoreceptor cell nuclei to the loss of nearly all retinal layers. RD must be distinguished from peripheral retinal degeneration,

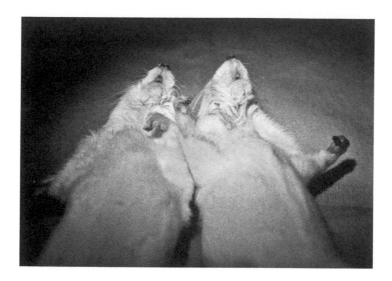

Fig. 52 Heat prostration can occur in rats shipped in hot or cold weather. Signs include increased salivation and wet muzzles and paws as the rats attempt to cool themselves.

which occurs as a genetically inherited disorder in some rat strains. High-intensity light exposure can induce glandular cell necrosis in the Harderian gland.

Adynamic Ileus Associated with Chloral Hydrate

Severe intestinal tract dilatation and ileus is associated with intraperitoneal injection of the anesthetic chloral hydrate. The abdomen is distended with dilated loops of atonic bowel. Lesions may require several weeks to develop following injection and do not develop in all rats injected. The lesions resemble Tyzzer's disease.

Pododermatitis

Also called bumblefoot, this is an inflammation and infection of the plantar surface and connective tissue of the foot (**Figure 53**). Lesions results from abraded tissue or pressure and begin as a small, reddish, raised area of keratinized growth that develops into crusts or scabs. There may be intermittent bleeding, secondary bacterial and fungal infections, and abscessation. Predisposing factors include long-term housing on wire-bottom cages, exercise on circular

Fig. 53 Pododermatitis, or bumblefoot, is more common in rats older than one year of age housed on wire-bottom cages. (Image courtesy Department of Comparative Medicine, College of Medicine, Penn State Hershey Medical Center.)

activity wheels causing abrasions and lacerations, and overweight rats placing continued or excessive pressure on the feet.

Auricular Chondritis

Auricular chondritis, or auricular chondropathy, is characterized by nodular to diffuse, auricular thickening in aged rats. Histologically, there is focal to multifocal granulomatous inflammation with cartilaginous plate destruction accompanying cartilaginous nodules and osseous metaplasia. The lesions reportedly occur spontaneously in some strains. Metallic ear tags have also been implicated, with milder but similar lesions seen on the untagged ear pinna of young SD rats (**Figure 54**). Similar lesions have been experimentally induced in type II collagen immunized rats, suggesting an immunologic component to the pathogenesis.

disease prevention, treatment, and control

A comprehensive discussion of treatment regimens for all rat diseases is beyond the scope of this text. **Table 16** lists select rat anti-infectives. The Hawk, Leary, and Morris (2005) *Formulary for Laboratory Animals* serves as an excellent reference.

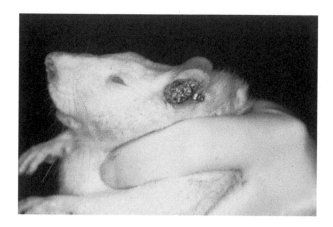

Fig. 54 Auricular chondritis associated with ear tags. This lesion is common in Wistar and SD rats. (Image courtesy Department of Comparative Medicine, College of Medicine, Penn State Hershey Medical Center.)

TABLE 16: COMMONLY USED RAT ANTI-INFECTIVES

Anti-infective	Dose
Amikacin	2–5 mg/kg SQ, IM q8-12h
Amoxicillin	150 mg/kg SQ, IM q12h
Ampicillin	50–150 mg/kg SQ q12h; 50 mg/kg IP q24h for 10d
Cephalexin	60 mg/kg PO q12h; 15 mg/kg SQ q12h
Enrofloxacin	10 mg/kg SQ q12h; 2.5–5 mg/kg PO, SQ, IM q12h
Fenbendazole	8 mg/kg BW (150 ppm) in medicated diet; feed for 7d followed by 7d on regular unmedicated diet; repeat cycle at least 3 × (minimum 42d)
Gentamicin	5–8 mg/kg SQ q24h
Griseofulvin	25 mg/100 g PO q24h for 10d
Ivermectin	200 µg/kg sid for 5d gastric gavage; 3 mg/kg PO once; or 2 mg/kg PO for 3 treatments at 7–9d intervals; diet, 12 ppm
Ketoconazole	10–40 mg/kg/d PO for 14d
Metronidazole	20–60 mg/kg q8-12h PO
Neomycin	2 mg/mL drinking water; 50 mg/kg IM q12h
Oxytetracycline	60 mg/kg SQ q72h of long-acting drug (Liquamycin® LA-200®)
Penicillin G	100,000 IU/kg IM q12h; 15,000 IU/20 mL drinking water
Piperazine	2 mg/mL drinking water
Sulfamethazine	12.5%: add 11 mL/L drinking water (100 mg/kg PO); 665–950 mg/L drinking water
Tetracycline	450–643 mg/L drinking water; 20 mg/kg PO q12h; 100 mg/kg SQ
Trimethoprim-sulfadiazine	0.5 mL/kg SQ of 240 mg/mL solution
Tylosin	10 mg/kg SQ q24h; 5 g/L in drinking water mixed with dextrose (give 100 mL treated water to each rat daily)

General disease treatment and control considerations:

- In rodent colonies, an ounce of prevention is truly worth a pound of cure, especially concerning pathogens. Thus, quarantine and sentinel programs are essential aspects of an effective veterinary care program. Additionally, strict barrier conditions and sound husbandry practices are important.

- Clinical signs are rarely specific to one disease. Any ill rat may have a roughened hair coat, hunched posture, porphyrin staining around the eyes and nose, and weight loss. Therefore, definitive disease diagnosis in the colony often depends on a thorough clinical and laboratory diagnostic workup and may include a complete necropsy of an affected animal(s) when feasible. Contemporary molecular techniques greatly facilitate rodent pathogen identification.

- Several procedures and factors require consideration when handling disease outbreaks, with the order and priority being unique to each facility and to the agent identified. These include determining the infection's origin and the agent's transmission mode(s). Infectious agents can be introduced into colonies by various means, including incoming rodent shipments, unscreened biologics, and fomites (e.g., people, contaminated food/bedding, other supplies). Further investigation can determine the pathogen's prevalence. The disease's effects on ongoing research studies and on all other colonies in the facility warrant evaluation. Lastly, the best ways to contain the outbreak and minimize its research effects in the facility need to be identified. Treatment, if applicable, includes removal from the study, isolation and quarantine, test-and-cull strategy, "burn-out," rederivation, depopulation followed by repopulation, or a combination of these methods tailored to the specific colony and research project. During disease resolution, close communication must be maintained within and between the research, veterinary, and husbandry groups.

- Treatment, although possible, may not be recommended for some infections. For example, salmonellosis treatment, while available, is not recommended, because a chronic carrier state may result and there is a potential for zoonotic disease. Some medications—especially antibiotics—may well ameliorate clinical signs but do not eliminate the disease from the population.

- Because few instances warrant treatment, most affected rats are euthanized rather than treated as the disease and/or treatment may alter the research paradigm. Treatment can be initiated if the condition is treatable, the animals are valuable to the study, and the resulting data would not be compromised by the treatment/condition. Before initiating any treatment, both the veterinarian and the principal investigator must discuss various items, including treatment options, treatment plans (including duration), and consequences to the animals and the associated research study. Genetically modified animals may require more frequent observation for treatment side-effects, including animal deaths, throughout the treatment's course. Give strong consideration to an initial test treatment of affected animals, where each sex, stock, strain, and genetic construct is treated for at least a 7-day observation period. A positive outcome of a test treatment does not guarantee similar results across the remaining animals.

- When treatment is necessary, two primary drug choice considerations are the drug's intended use and its safety, as highlighted above. Drug choice is first determined by the drug's spectrum of activity, strength (potency), and effectiveness (efficacy). For example, penicillin-based antibiotics are used for Gram-positive bacterial infections. Possible effects on the animal's physiology and the study are then identified. For example, ivermectin, a common parasite treatment, may be lethal in animals (certain stocks, strains, genetic constructs and/or neonates) with an incomplete blood-brain barrier. Meanwhile, the antibiotic gentamicin is ototoxic and nephrotoxic, making it a poor choice for animals used in hearing or kidney research.

- A combination of various drugs and treatment modalities may be required to complete a treatment regimen. Clinical signs and symptoms can be eliminated or at least alleviated by drugs. Examples are anti-inflammatory medications and bronchodilators. Fluid therapy is necessary to correct an animal's hydration status. A recirculating warm water blanket, among other things, can minimize hypothermia. However, if at all possible, the primary goal is to eliminate the cause of the clinical disease. For example, fenbendazole may be used to eliminate a pinworm infection. For traumatic injuries, the cause of the injury can be removed and/or the animal's environment can be enhanced.

- When administering oral medications, the preparation's palatability is important, necessitating the monitoring of food and water intake. Some flavored preparations are available commercially (e.g., bubblegum-flavored amoxicillin, bacon-flavored carprofen, and bacon-flavored "*Helicobacter* diet"). Alternatively, drugs can be made palatable by mixing and coating with compatible substances such as peanut butter and human nutrition shakes. Regular diet and water can also be made more palatable for ill rats.

- Morris (1995) discusses issues concerning antibiotic use in laboratory animals.

Bacterial and Fungal Diseases

- Implement antibiotic and antifungal therapy based on culture and sensitivity testing. This is especially true for bacterial species having different strains and for infections refractory to common treatment.

- Infected bite, fight wounds, and other superficial skin injuries can be treated locally by cleaning the site and applying a topical antibiotic.

- Ulcerative dermatitis caused by *Staphylococcus aureus* may be treated with local or systemic antibiotics if the initiating factors are removed. Clipping the hind toenails may reduce lesion severity.

- *Pseudomonas* infections can be difficult to control. Many isolates are resistant to a variety of antibiotics. Phenolic disinfectants usually are effective, but organisms can thrive in quaternary ammonium compounds and iodine solutions. Hyperchlorinating (10–12 ppm) or acidifying (pH of 2.3 to 2.5) drinking water reduces the incidence of clinical disease.

- Three-drug (amoxicillin, metronidazole, and bismuth) and four-drug (omeprazole, metronidazole, amoxicillin, and clarithromycin) combinations have successfully eradicated rodent *Helicobacter* infections. Commercially available feed with either combination is available to overcome the labor-intensive nature of these regimens. Concern exists that *Helicobacter* may not be eradicated and may recur following antibiotic removal. This has driven some institutions to pursue the more successful and rewarding rederivation (e.g., cross fostering) over treatment. Antibiotic treatment may be used to decrease bacterial shedding prior to cross fostering.

- *Pneumocystis carinii* treatment and prevention in immunocompromised rats is by oral administration of trimethoprim-sulfamethoxazole. However, a sulfonamide alone might provide effective prophylaxis. The authors do not recommend colony maintenance on sulfonamides in the absence of infection, as antibiotics use must not replace sound animal handling and husbandry practices.

Viral Diseases

- Currently, there are no rat viral disease treatments. If necessary, clinical signs may be alleviated, but this comes with the risk of exposing the remaining animals in the facility to the infection.

- A test-and-cull strategy may be sufficient in eradicating a viral infection from a colony, as in the case of rat theilovirus.

- Where transplacental transmission is not significant, rederivation is recommended.

- If rederivation is not possible, breeder isolation and repopulation with seronegative young may control the disease. Quarantine the room for at least 6 weeks, during which time all breeding is stopped, suckling and weanling animals are euthanized, no new animals are introduced, and steps taken to prevent cross contamination (e.g., through shared lab equipment). Following quarantine, place sentinels in the room and test them 3–4 weeks later to ensure active disease is not present.

- One may also perform "burn-out." This consists of preventing new colony introductions, ceasing breeding (6 weeks minimum), and deliberately spreading the infection (e.g., dirty bedding, removing filter tops) to encourage global colony disease spread; "burn-out" failure is common.

Parasitic Diseases

- Lice and mites can be treated with a variety of drugs, including pyrethrins, carbaryl, and topical avermectins (e.g., ivermectin, selamectin). Ivermectin-compounded feed is available and has successfully treated mouse fur mites.

Ivermectin has been used successfully in rat pups, but should be avoided if possible because ivermectin accumulates in the dam's milk, potentially exposing pups to toxic levels. Furthermore, pups up to 10 days of age have an incomplete blood–brain barrier.

- Pritchett and Johnston (2002) reviewed rodent colony pinworm treatments. Piperazine in the drinking water, fenbendazole-medicated feed, oral dosing with ivermectin, and dietary doramectin and netobimin were effective in eliminating pinworm infections. The use of topical avermectins and dietary-administered moxidectin were less successful. Control involves rigid sanitation and using filter-top cages. The eggs are extremely hardy, very light, and may aerosolize. Physical methods such as scrubbing with detergent and steam cleaning are perhaps the most effective for environmental decontamination. In one study, formaldehyde gas and chlorine dioxide treatments showed marked effectiveness against *Syphacia muris* eggs.

- Long-term parasite control relies on continued rat colony monitoring because the parasite eggs are well suited to survive within the environment.

Neoplasia

- Mammary fibroadenoma is usually benign and amenable to surgical removal; however, new tumors may develop anywhere along the line of mammary tissue.
- Caloric restriction will reduce the incidence of neoplasia.

Age-Related Lesions

Ad libitum diet influences the development of age-related lesions. For example, restricting calories and protein content reduces the occurrence of chronic progressive nephropathy (CPN). Caloric restriction may prolong lifespan; further discussion is found in Chapter 2.

Miscellaneous Conditions/Diseases

- Malocclusion is best treated using a dental drill or small grinder to evenly grind the teeth to their appropriate size. Clipping the teeth can cause splitting and the development of tooth root abscess. Trimming needs to be routinely repeated and can be as often as every 2–4 weeks. Powdered rat diet can be added to the pelleted diet; in case rats cannot use their incisors adequately to remove and gnaw pellets, they could use their molars to chew the powdered diet.

- Pododermatitis treatment includes cleaning lesions with an antiseptic solution and the application of a topical antibiotic ointment. Correction of the underlying problem (e.g., changing to solid-bottom cages) is warranted. Administration of an oral broad-spectrum antibiotic, surgical debridement, and, in advanced cases, amputation of the affected extremity or euthanasia may also be needed.

- Treat heat stress with careful spraying with a gentle stream of cool water and by subcutaneous fluids. Residual heat stress effects may make recovered animals unsuitable research subjects. Prevent heat stress, for example, by appropriate use of environmentally controlled vehicles (e.g., avoid placing animals near heater vents) and by careful weather monitoring prior to shipment.

- Auricular chondritis, related to using ear tags, can be prevented by using other identification methods such as ear punching.

detecting and preventing infectious disease

Rodent colonies are generally treated on a "herd-health" basis, that is, prevention, treatment, and control measures are at the colony level rather than the individual level. Disease prevention and identification measures take the form of three major health status programs:

1. Quarantining incoming animals and/or vendor surveillance, whereby the health status of animals arriving from various sources is determined prior to their integration to the existing colony.

2. Screening biologics (e.g., serum and animal and human tumor and cell lines) for pathogens.

3. Colony health surveillance using a "true" sentinel animal program. Random or selective colony sampling and/or direct-contact sentinel exposure may be warranted for some

pathogens that cannot be reliably detected through soiled-bedding sentinel exposure.

Quarantining Incoming Rats/Vendor Surveillance

Incoming animals may be from established commercial vendors with stringent disease prevention strategies ("approved vendors") or sources with less stringent or nonexistent strategies ("unapproved vendors"). To minimize disease risk, it is advisable to process shipments from approved and unapproved vendors differently. Ideally, disease-free rats are acquired and maintained disease-free following their arrival in the facility.

Incoming rat shipments may be held in quarantine until the animals' health status is confirmed either through testing of a sample from the shipment itself (vendor surveillance) or placing sentinel rats to be tested at least 6 weeks following exposure to the new arrivals. Pathogen testing may be similar to colony sentinels; however, some institutions choose more comprehensive sentinel test panels. Some institutions elect to feed quarantine animals diets containing either fenbendazole or ivermectin. An alternative to quarantining animals is to rederive all animals entering the animal facility.

Screening Biologics

Biologics (e.g., cell lines, tumors, sera, antibodies) may be contaminated with murine pathogens, yielding active infections when inoculated into rodents, and/or with human pathogens, posing a health risk to personnel. It may be unnecessary to test biologics unexposed to rodent tissues or rodent-derived substrates for rodent pathogens; however, unless a detailed and complete history of the biologic is available, there may have been opportunities for the biologic to become contaminated. Meanwhile, human origin biologics, such as human cell lines and tumors, may be contaminated with human pathogens, thus posing health risks to personnel handling the specimens and can confound research results.

Biologic screening methodologies include PCR (human and rodent pathogens), viral plaque assays, and rat antibody production (RAP) testing, with PCR being categorically the most sensitive and rapid test. With RAP, pathogen-free test rats develop antibody titers detectable by serology 4 wk following inoculation of the contaminated biologic. Uninoculated sentinel rats can be used as test controls to rule out the possibility of a pre-existing colony infection.

Note: *Mycoplasma* spp. is a common contaminant of biologics, and nonrodent mycoplasmas are more common than *M. pulmonis*. Anecdotal evidence indicates some non-*M. pulmonis* mycoplasmas are responsible for animal deaths when inoculated into immunocompromised rodents.

Colony Health Surveillance

Appropriate diagnostic testing and institutional interpretation of results are important to ensure high-quality, healthy laboratory rats. Each institution must develop a pathogen exclusion list and an appropriate sentinel program. Sentinel programs provide a practical, valuable, economic, and logistically feasible alternative to testing actual research colony animals. These programs compliment sound husbandry practices and current engineering standards in preventing and identifying pathogens that may negatively impact research animals. Shek (2008), Compton et al. (2004), and Koszdin and DiGiacomo (2002) present thorough discussions of this topic.

Sentinel animal exposure to colony animals can either be by direct contact or dirty bedding. There are advantages and disadvantages of using either. Some pathogens can be accurately monitored only by using direct contact primarily owing to the agent's mode of transmission (e.g., direct contact seen with mites). Contemporary husbandry practices (e.g., autoclaved individually ventilated cages, hyperchlorinated/acidified water, HEPA-filtered cage change stations) may also hamper direct pathogen transmission or dilute contaminated bedding (reducing/eliminating the infectious dose) provided to sentinels. However, contact sentinel animals risk the colony's health status, as sentinel animals are potential sources of genetic and microbial contamination; furthermore, other negative scenarios may occur, such as fighting. Soiled bedding exposure is logistically simple and a more common method. Unlike direct contact sentinels, the effectiveness of bedding transfer protocols varies between facilities. One reason is that the quantity of an infectious agent found in bedding pooled from many cages can easily be diluted below an infectious dose to the sentinel animals. There is a wide array of exposure protocols (especially regarding testing frequency and bedding volume transferred) used by institutions. Soiled bedding sentinel exposure is reportedly unreliable in detecting Sendai virus, SDAV, LCMV, CAR bacillus, and fur mites.

Sentinel rats are commonly 4–6-week-old female SD rats from a known pathogen-free source. The testing frequency and the pathogens evaluated depend on the facility, strain, and colony dynamics (e.g., breeding colony, aging study). Most facilities screen animals quarterly. Screening usually occurs after at least 4 weeks to ensure adequate exposure and detectable antibody titer development. The actual number of sentinel animals usually depends on the room's population, disease incidence, and housing type. However, the goal remains to have quality animal care coupled with practical and affordable health monitoring.

Sentinel programs may contain all or a selection of the following examinations and tests:

- Gross necropsy and visual examination of organs for abnormalities; histopathology of representative tissues and/or grossly abnormal tissues/organs.
- Culture and PCR for both respiratory and enteric bacterial pathogens. This may include using specialized media for recovering agents such as *Salmonella* and *Mycoplasma*. PCR for *Helicobacter* spp. and *Pneumocystis* are available. Dermatophyte cultures are not routinely performed in the absence of skin lesions.
- Serologic assays (primarily MFI) are available for a number of pathogens. Hantavirus serologic testing is not routinely performed; however, animal shipments to some countries may request Hantavirus (or other exotic agent) testing.
- Endoparasite examination via fecal flotation, anal tape test, cecal contents, and/or PCR.
- Ectoparasite examination via pelt examination (fur pluck, dorsal tape test, skin scrape, direct microscopic exam of excised pelt), and/or PCR.

Other Aspects of Disease Prevention and Control

- Pest control program: Feral rats and mice, pet rodents, and other vermin (e.g., insects) may carry pathogens (e.g., *M. pulmonis*).
- Sanitation and decontamination: Includes animal housing (micro- and macroenvironment), procedure rooms, experimental equipment and instruments, laboratories, water and feed, and other items that may come in direct contact with the animals.
- Other practices, including properly using personal protective equipment, biosafety cabinets, cage change stations,

and animal housing modalities. Policies that involve such practices should also be implemented. For example, a facility training/orientation outlines expectations for all individuals (research, husbandry, and veterinary staff).

Rederivation

Rederivation may be necessary to eradicate various pathogens. It can also improve the general health and welfare of animals by eliminating or reducing the incidence of certain clinical entities. It can be accomplished by cesarean section (CS), embryo transfer (ET), or cross fostering (CF). Of the three methodologies, CF is a viable low-tech, low-cost method and is often done by disinfecting 24-hour-old neonates with iodine solution or 100-ppm chlorine dioxide and transferring them to pathogen-free foster dams. CS and ET require euthanasia of dams to obtain the pups or embryos, respectively. CS rederivation requires less equipment and resources than ET, although ET is widely regarded as the safer and more effective method. Detailed ET discussions are beyond the scope of this text, but the procedure can be conducted by most transgenic core facilities with proper precautions. It is advisable to consult the ET rederivation service provider. For further information, consult Pinkert (2002), Charreau et al., (1996), and Rouleau et al. (1993).

Cesarean section preparation and procedure

1. Determine the gestation period of the stock or strain of rats undergoing CS. This is especially a common obstacle with transgenic and knockout lines in which gestation length often varies. The gestation period is typically measured from the day the vaginal plug is seen or the vaginal smear contains sperm. A few test animals should go through full gestation to determine the specific length of gestation.

2. The CS should occur as close to the time of birth as possible. By observing the animals frequently, one can determine whether the CS should occur in the morning or afternoon. A good schedule is to check the test animals at 7 a.m., 11 a.m., 1 p.m., and 6 p.m. If the animals typically deliver by 6 PM, one should perform the CS by late morning or early afternoon.

3. At least an equal number of foster dams should be available. Plan for these foster mothers to deliver the day before the CS is to occur.

4. The room setup should include

   a. A separate area where the contaminated or "dirty" mothers are euthanized (decapitation) and the gravid uterine horns are removed. This is the "dirty" or contaminated area.

   b. A warm (80–85°F, 26.7–29.4°C) room

   c. Warm iodine or povidone-iodine solution

   d. Warm saline

   e. Heating pads, with chemical heating pouches being very effective for maintaining the neonates' body temperature during transport

   f. Sterile cotton swabs

   g. Instruments:
      i. scissors
      ii. mosquito forceps (3)
      iii. tissue forceps

   h. Cotton sponges or Telfa® pads

   i. Doxapram hydrochloride

   j. Petri dishes

   k. Towels

   l. At least two people: one working with the clean animals and one working with the contaminated animals

Perform the procedures as quickly and as safely as possible. The delivery of rats via CS and experiencing a period of less than five minutes of anoxia does not appear to have effects such as spatial learning deficits. The procedure is as follows:

1. Administer doxapram hydrochloride 5–10 mg/kg SQ to the contaminated dam; wait 5 min.

2. Euthanize the contaminated dam. Physical euthanasia methods, such as decapitation, are preferred to pharmacologic methods (e.g., pentobarbital), as pharmacologic methods can result in cardiopulmonary depression in the pups.

3. Prepare the area by wetting the fur with alcohol.

4. Remove the gravid uterus through a midline incision, and clamp off the two distal horns of the uterus and the cervix. Cut the necessary attachments, being careful not to enter the bowel.

5. While holding onto the three forceps, dunk the gravid uterus into the warm iodine solution.

6. Move to the clean area.

7. Place the iodine-soaked gravid uterus onto the clean area and remove the "dirty" forceps. Return the "dirty" forceps to the contaminated area.

8. Quickly remove the fetuses from the uterus along with their fetal membranes. Place the fetuses onto the sterile Petri dishes lined with Telfa pads or cotton sponges. The Petri dishes should be atop a heating pad. If using a chemical heat pouch, there should be one layer of the towel between the heat pouch and the Petri dish.

9. Stimulate the neonates by rolling them around with the cotton-tipped swabs. To prevent them from sticking to the cotton sponges, the sponges may be moistened with warm saline. The animals should not stick to the Telfa pads. Moistening the cotton sponges acts to remove amniotic fluid and blood, which, if remaining, could result in cannibalization of the young.

10. Exclude any young that do not appear to be responding, to minimize the risk of cannibalization.

11. Transport the animals to the foster, "clean" mother. Take care to keep them warm using the chemical heat pouches and towels during transport.

12. Ensure the total number of pups (foster + natural-born pups) approximates or equals the original litter size of the foster mother's cage. To ensure fostering one may wish to have the foster mother urinate on the foster pups. The foster mother's pups are typically one to two days of age at the time of fostering. The removed foster mother's pups are then euthanized.

13. The foster mother should remain undisturbed for at least 48 hr.

---

anesthesia and analgesia

One of the most important aspects of laboratory animal medicine is providing appropriate analgesia and anesthesia. It requires professional veterinary judgment to properly implement such regimens and avoid toxicity or possible interference with experimental data.

There is much terminology concerning anesthesia, analgesia, and sedation. Below is a list of commonly used terms.

Akinesia: The loss of motor responses.

Analgesia: The absence of pain in response to stimulation that would normally be painful. An agent that temporarily reduces or eliminates the sensation of pain is called an *analgesic*.

Anesthesia: A temporary and reversible state characterized by a marked reduction or elimination of sensory and motor responses.

Balanced anesthesia: Producing general anesthesia by combining two or more drugs or anesthetic regimens.

Light anesthesia: Animals under light anesthesia are immobilized and have lost the ability to right themselves; they will react to some painful stimuli, however.

Local anesthesia: Desensitization of specific, defined sites of the body, often involving the skin and subcutaneous tissues by inhibiting excitation of nerve endings or by blocking conduction in peripheral nerves.

Neuroleptanalgesia: A trance-like state accompanied by analgesia seen with the use of sedation and analgesic agents.

Noxious stimuli: Events that damage or threaten damage to tissues (e.g., cutting, crushing, or burning stimuli) and that activate specialized sensory nerve endings called nociceptors.

Pain: A sensory response evoked by a noxious stimulus; in humans, pain is "an unpleasant sensory and emotional experience associated with actual or potential tissue damage, or described in terms of such damage" (International Association for the Study of Pain).

Sedation/tranquilization: Mild CNS depression where the patient is awake and calm.

Surgical (General) anesthesia: Characterized by the loss of consciousness and protective reflexes, analgesia, and muscle relaxation permitting the performance of a surgical procedure without patient pain or movement.

With the assistance of a veterinarian or veterinary technician, evaluate the points listed below to determine whether local anesthesia, general anesthesia, or sedation and what anesthetic and analgesic regimens are appropriate for a given procedure. Generally, balanced anesthesia regimens are optimal.

- Purpose: Study or procedure to be performed (e.g., physical examination, surgery, body fluid collection); its invasiveness and pain potential; and experimental constraints (e.g., bioimaging, laparotomy, cardiothoracic surgery).

- Signalment and animal health status: Strain, age, size, sex, temperament, physical condition, disease, physiologic status, and stressors can yield different analgesic and anesthetic responses. Ill patients, geriatrics and neonates, obese, emaciated, and pregnant rats need anesthetic agents and techniques that consider, yet not compromise and/or exacerbate, the physiologic consequences of their conditions. Younger male SD rats, as young as 23 days, reportedly require larger propofol induction doses than older counterparts. Younger rats tend to have a higher inhalant anesthesia minimum alveolar concentration requirement than older animals. In neonatal rats, methoxyflurane and hypothermia were safer and more effective than ketamine, pentobarbital, and fentanyl-droperidol. Stressors, including early maternal separation, have lasting effects on nociception and analgesic efficacy; morphine is reportedly less effective in cross-fostered rats. Strain differences include the strong ACI rat response to buprenorphine, the relative resistance of F344 and BN rats to medetomidine, the relative hypersensitivity of the ACI and BN rats to ketamine, and sodium pentobarbital's significantly lower lethal dose (LD_{50}) in several albino rat strains compared to pigmented rats.

- Postprocedural needs/manipulations: Postprocedural analgesia, sedation, and/or anesthesia.

- Others: Equipment and agent availability, training and expertise of the anesthetist, specialized equipment (e.g., MRI-friendly anesthesia and anesthesia monitoring equipment).

Analgesia

Animal pain and distress management is complex, and although it is difficult to measure pain, particularly in rodents, progress in understanding its dynamics and mechanisms, assessment, and effective treatment is being made. For example, reports suggest rats lack opioid receptors in their cerebellum and cerebral cortex, and that rat pups aged 5–7 days and younger are not likely to perceive pain, with the ability to perceive pain developing gradually between postnatal ages 12–14 and 21–22 days. Arthritic rats reportedly learn to self-administer or self-select food or water containing analgesic drugs. Pain-scoring systems based on various clinical manifestations of pain such as water consumption and stretching and back-arching behavior, and guidelines for assessment and diagnosis of analgesic needs are available. Lastly, different strategies such as nonpharmacological considerations (e.g., modified housing and husbandry practices, dietary modifications), pharmacological interventions, or euthanasia may be employed. One of the most popular principles behind animal analgesia is the anthropomorphic view: that is, if the proposed experimental procedure were performed on a human instead, then the same level of pain control required for that person to comfortably survive and recover from the procedure should be automatically provided to the animal. The commentary from Gross et al. (2003) elaborates on this topic, while recommendations for improving rodent pain management based on existing literature are reported by Stokes et al. (2009).

There are six broad groups of drugs used in pain management: opioids (**Table 17**), nonsteroidal anti-inflammatory drugs (NSAIDs) (**Table 18**), local anesthetics, α_2-adrenergic agonists, N-methyl-d-aspartate (NMDA)-receptor antagonists, and analgesic adjuncts (**Table 19**).

Opioids

Centrally and peripherally acting compounds that work in conjunction with the endogenous opioid receptors (mu, kappa, and delta) of the nervous system (and other tissues). The most effective opioid analgesics are associated with mu-receptor activity.

NSAIDs

These analgesic, antipyretic, and anti-inflammatory agents inhibit cyclooxygenase (COX)-1, -2, and -3 isoenzymes, catalyzing the formation of inflammatory modulators. Rats appear more susceptible than other rodents to gastrointestinal ulcers resulting from the use of certain NSAIDs. Exposing dams at 9 days of gestation to NSAIDs that selectively inhibit COX-1 or that have a high ratio of COX-1 to COX-2 inhibition (i.e., ibuprofen, diflunisal, and ketorolac) was linked to fetal-heart developmental anomalies (ventricular septal defects and midline defects).

TABLE **17**: **OPIOID ANALGESICS**

Agent	Receptor Activity	Recommended Dose
Buprenorphine[a]	Partial mu agonist; kappa and delta antagonist	0.01–0.05 mg/kg IV, SQ q8-12hr
Buprenorphine SR™[b]	Partial mu agonist; kappa and delta antagonist	1.2 mg/kg q72hr
Butorphanol	Partial mu antagonist/agonist; kappa agonist	2 mg/kg SQ q4hr
Meperidine	Kappa agonist	10–20 mg/kg IM, SQ q2-3hr
Morphine	Mu agonist	10 mg/kg SQ q2-4hr
Nalbuphine	Partial mu antagonist; kappa agonist	1–2 mg/kg IM q3hr
Pentazocine	(–) enantiomer: Kappa agonist; (+) enantiomer: Delta agonist	5–10 mg/kg SQ, IM q3-4hr
Oxymorphone[c]	Mu agonist	0.15 mg/kg IM
Tramadol	Mu agonist	10–12.5 mg/kg IP q12-24hr

[a] Buprenorphine is the most commonly used analgesic in small mammals such as rats. Small animals' disproportionately faster rates of drug metabolism than larger species can impact buprenorphine's duration of action. High doses may have a slight immunostimulatory effect and may result in decreased urine reduction and pica, the latter leading to gastric distention. Like butorphanol, buprenorphine reportedly causes reduced food intake and increased locomotor activity postoperatively in rats. Buprenorphine has anesthesia-sparing effects such that if given preoperatively, the isoflurane concentration can be reduced by 0.25–0.50%, and unpredictable increases in anesthetic depth by injectable anesthetics (e.g., ketamine/xylazine) may occur. Buprenorphine requires higher doses when given orally as it undergoes significant first-pass metabolism in the liver with this route.

[b] Buprenorphine sustained-release.

[c] Oxymorphone is reportedly a superior analgesic to buprenorphine for alleviating visceral pain due to intestinal resection in rats. Rats can be treated with continuous IV infusion (0.03 mg/kg/hr) or by bolus IV injection.

Local anesthetics

Local anesthetics (LAs) are reversible membrane stabilizing agents generally acting in the area they are instilled. The duration of action depends on the specific drug used or on combining a local anesthetic with a vasoconstricting agent such as epinephrine.

In rats, LAs can be placed directly into or around the site of interest. Only a small LA volume may be necessary for desensitization; however, LAs in rodents appear to have a shorter duration of action than in larger animal species. The two most commonly used LAs are bupivacaine and lidocaine; lidocaine has faster onset than bupivacaine but shorter duration of action. Injectable solutions, ointments and creams (e.g., lignocaine-prilocaine cream), and sprays (e.g., xylocaine spray) are available. Epinephrine may

TABLE 18: **NSAID ANALGESICS**

Agent	Mechanism of Action (Inhibits)	Recommended Dose
Acetaminophen	COX-2	100–300 mg/kg PO q4hr; 6 mg/mL drinking water
Aspirin[a]	COX-1 and -2	100 mg/kg PO q4hr
Carprofen	COX-2	2–5 mg/kg SQ q24hr
Diclofenac	COX-1 and -2	10 mg/kg PO q24hr
Flunixin	COX-1 and -2	2.5 mg/kg IM, SQ q12-24hr
Ibuprofen	COX-1 and -2	15 mg/kg PO q24hr
Ketoprofen[b]	COX-1 and -2	5 mg/kg SQ q24hr
Meloxicam	COX-2	1 mg/kg PO, SQ q24hr
Piroxicam	COX-1 and -2	3 mg/kg PO q24hr

[a] High aspirin doses produce developmental anomalies in rats (e.g., umbilical and diaphragmatic hernia) when administered during sensitive windows of development.
[b] Ketoprofen can cause gastrointestinal toxicosis in rats at therapeutic doses.

TABLE 19: **ANALGESIC ADJUNCTS**

Agent	Mechanism of Action/Compound	Dose
Amantadine*	NMDA receptor antagonist, antiviral, Parkinson's disease treatment	25–50 mg/kg IP
Amitriptyline	Antidepressant	20 mg/kg IP
Amoxicillin	Antibiotic	100–300 mg/kg SQ, IP
Ampicillin	Antibiotic	100 mg/kg IP
Ciprofloxacin	Antibiotic	32 mg/kg IP
Gabapentin	γ-aminobutyric acid structural analog; anticonvulsant	80 mg/kg q24h IP
Gentamicin	Antibiotic	4 mg/kg IP
Verapamil	Calcium channel blocking agent	5 mg/kg IP
$MgSO_4$*	NMDA receptor antagonist	60 µg/h intrathecally
Memantine*	NMDA receptor antagonist; Alzheimer's disease treatment	10 mg/kg IP

* Synergy is reported between morphine (µ agonist) and these NMDA receptor antagonists.

be added to the LA solution to improve safety, decrease the LA dose, and prolong the duration of action by delaying LA absorption. LA administration carries the potential hazard of IV injection, subsequently inducing life-threatening CNS and cardiovascular toxicity. This underscores the importance of aspiration prior to administering any injectable compound. In the rat, and compared to mepivacaine and prilocaine, intrathecal procaine and lidocaine are the least and most neurotoxic, respectively. Meanwhile, levobupivacaine and ropivacaine are two new long-acting LAs less neurotoxic than bupivacaine. The maximum safe doses are 4 mg/kg lidocaine (0.4 mL/kg of 1% solution) and 1–2 mg/kg bupivacaine (0.4–0.8 mL/kg of 0.25% solution).

α_2-adrenergic agonists

These compounds produce varying levels of sedation, analgesia, muscle relaxation, and anxiolysis, and decrease the injectable and inhalant anesthetic agent requirements (dose sparing). Examples are xylazine, dexmedetomidine, and medetomidine.

N-methyl-d-aspartate (NMDA)-receptor antagonists

These compounds inhibit NMDA receptors' action, producing an anesthetic state referred to as dissociative anesthesia. Examples are ketamine, amantadine, magnesium, and dextromethorphan.

Adjunctive agents

While these drugs may not have analgesic activity alone, they may act with analgesics to benefit pain management. Examples are gabapentin, antibiotics, calcium channel blockers, and tricyclic antidepressants.

Preoperative Medications

Preanesthetic medications facilitate handling, reduce the dose of induction and/or maintenance anesthetic agents, minimize fluctuations in the level of anesthesia, provide preemptive analgesia and antibiotic prophylaxis, and minimize anesthetic agent side-effects. Preemptive oral analgesia is better than postoperative oral analgesia to ensure analgesic consumption, as pain may depress postoperative food intake. Preemptive oral analgesia may start as soon as 2 days prior to the procedure to also acclimatize the rats to the novel water/food, ensuring adequate postoperative intake. Most rodents seem to benefit from a single injection of buprenorphine intraoperatively or immediately postoperatively; however, it is imperative to administer analgesia prior to the noxious insult (e.g., surgery) and continue analgesia for a suitable period afterwards. Preoperative NSAID use (except carprofen) should be avoided because of the potential for renal toxicity should hypotension occur during anesthesia. For antibiotics, only a single bolus is usually needed unless there is great likelihood of postoperative infection (e.g., gastrointestinal tract procedures, contaminated wound). Other preanesthetic agents may include atropine or glycopyrrolate. Either is used to decrease salivation and to counter decreased heart rate caused by increased vagal tone, but atropine has a faster onset and shorter duration of action than glycopyrrolate and is recommended for cardiac emergencies (bradycardia). In rats, these drugs are often given at anesthetic induction to avoid repeated handling and stress; their use introduces the likelihood that saliva will be more viscous and presents additional patient management concerns. Other agents are used for sedation and immobilization.

Table 20 lists select preanesthetic agents, their dosages, and effects. Prior to administering anesthetic agents, it is recommended to allow sufficient animal acclimation following arrival in the facility and before experimental use. Fasting is generally not necessary, since rats are unable to vomit; if fasting is necessary, limit the fasting time to two hours or less because of the rat's high basal metabolic rate.

Anesthesia

Anesthetic agents may be delivered via injection or inhalation. When using inhalation agents, it is important to determine whether the animal has a respiratory tract illness, as this may interfere with

TABLE 20: SELECT PREANESTHETIC AGENTS

Agent	Dose	Effects
Acepromazine	1–2 mg/kg IM	Sedation
Atropine	0.05 mg/kg IP, SQ	Parasympatholytic
Diazepam	2.5–4.0 mg/kg IM, IP	Sedation
Dexmedetomidine + Midazolam	0.06–0.12 mg/kg + 10–20 mg/kg	Sedation/analgesia
Fentanyl/Droperidol (Innovar-Vet®)	0.5 mL/kg IM	Immobilization/analgesia
Fentanyl/Fluanisone (Hypnorm®)[a]	0.3–0.5 mL/kg IP, SQ; 0.3–0.6 mL/kg IP	Sedation/some analgesia
Glycopyrrolate	0.5 mg/kg IM	Parasympatholytic
Ketamine	50–100 mg/kg IM, IP	Sedation, Immobilization
Medetomidine	0.03–0.1 mg/kg IP, SQ	Sedation/some analgesia
Midazolam	5 mg/kg IP	Sedation
Xylazine[b]	4–8 mg/kg IP	Sedation/some analgesia

[a] Hypnorm® is currently unavailable in the United States. Its use in unfasted rats may result in hyperglycemia, possibly because of opioid involvement in glucose metabolism.

[b] α_2 agonists such as xylazine can cause hyperglycemia, and pulmonary edema at higher dosages (>20 mg/kg); some rats are susceptible to pulmonary edema even at published anesthetic doses. Care must be taken to use the proper concentration since xylazine is available in two different concentrations (20 mg/mL and 100 mg/mL). Xylazine also appears to affect antidiuretic hormone (ADH) secretion and glucose homeostasis.

anesthetic delivery, and ensure that the facility has appropriate gas-scavenging abilities as determined by the institution's occupational health and safety professionals.

Injectable anesthesia

If waste anesthetic gas-scavenging equipment is unavailable, it may be necessary to rely solely upon injectable anesthetic agents. Balanced anesthesia is common when working with injectable agents. Thus, it is important to consider accurate dosing with correct multidrug-use ratios, storage conditions, and the feasibility of immediate use following (re)constitution. Aside from potency loss from prolonged anesthetic cocktail storage, bacteria, endotoxin, and other contaminants are major concerns. Accurately weighing each animal immediately prior to administering a calculated anesthetic dose is critical to avoid either over- or underdosing. **Tables 21** and **22** provide information about agents, dosages, anesthesia time, and sleep time of common injectable anesthetics. Anesthesia time is the time one expects the rat to have a marked reduction or elimination of sensory and motor responses. Sleep time is the time one expects the rat to be unconscious. During sleep time, the anesthetic's effect can gradually diminish, resulting in responses to painful stimuli and thereby necessitating an additional anesthetic dose.

There are several advantages and disadvantages of using injectable anesthetics. One advantage of some injectable agents is the availability of injectable antagonist or reversal agents (**Table 23**). One may choose to use an antagonist if the animal has an adverse anesthetic reaction, receives too much anesthetic, or if the procedure is complete. Before administering reversal agents at the end of a painful procedure, administration of another suitable analgesic agent may be necessary. One disadvantage is that the dosage cannot be reduced after induction. Thus, drugs at either a low dose or with a wide safety margin should be used. Another disadvantage is the need for intermittent injections for anesthetic maintenance, thereby interfering and interrupting a procedure and potentially predisposing the animal to toxicity. Syringe pumps for continuous IV administration (i.e., infusion) provide more reliable, consistent, and accurate anesthetic dosing than multiple manual injections. Target-controlled infusion (TCI), a method using a computer-controlled infusion pump to deliver a drug based on a mathematically predicted plasma or "effect site" (e.g., brain for general anesthesia) concentration, can also be used.

TABLE 21: INJECTABLE AGENTS SUITABLE FOR SURGICAL ANESTHESIA

Anesthesia Time (min)	Agent(s)	Dose	Sleep Time (min)*
≤5	Alphaxalone/Alphadolone (Althesin®, Saffan®)	10–12 mg/kg IV; 0.2–0.7 mg/kg/min IV	10
	Methohexital	10–15 mg/kg IV	10
	Propofol[a]	10–20 mg/kg IV 0.5–1.0 mg/kg/min (infusion)	10
≤10	Thiopental	30 mg/kg IV	15
≥20	Fentanyl/Fluanisone (Hypnorm®) + Diazepam	0.6 mL/kg + 2.5 mg/kg IP	120–240
	Fentanyl/Fluanisone (Hypnorm®) + Midazolam[b]	2.7 mL/kg IP	120–240
	Ketamine + Dexmedetomidine[c]	75mg/kg + 0.5 mg/kg	120–240
	Ketamine + Medetomidine[d]	75 mg/kg + 0.5 mg/kg IP	120–240
	Ketamine + Xylazine	40–80 mg/kg + 5–10 mg/kg IP; 37 mg/kg + 7 mg/kg IM followed by 1–1.25 mg/kg/min IV + 32–40 µg/kg/min IV for up to 12 hr (infusion)	120–240
	Tribromoethanol	300 mg/kg IP	—

≥60	Fentanyl + Medetomidine[e]	0.3 mg/kg + 0.2–0.3 mg/kg IP	240–360
	Inactin (ethylmalonyl urea, or EMTU)	80 mg/kg IP	60–240
≥360	Urethane[f]	1000–1200 mg/kg IP	360–480

[a] Propofol at high infusion rates may produce apnea and extended recovery times; therefore it may be a better decision to use propofol with Hypnorm® (VetaPharma Ltd, Sherburn-in-Elmet, Leeds, UK), as this combination requires lower infusion rates and provides adequate analgesia, and the Hypnorm® is reversible. Premedicate with 0.5–1.0 mL/kg and infuse propofol continuously (4–6 mL/kg/hr) to provide stress-free induction, well-controlled anesthesia, good analgesia, and muscle relaxation. Meanwhile, buprenorphine administration at the start of periodic IV propofol anesthesia in rats has been reported to result in significant reduction in the total propofol requirement and improvement in recovery. Ventilatory support is strongly recommended for propofol anesthesia. Rapid bolus administration over less than 5 sec results in excitatory responses such as forelimb extensor rigidity and vibrissae twitching. In another study, propofol administration IV for 3 min increased the anesthetic duration and decreased cardiopulmonary depression in rats.

[b] Cocktail preparation: Combine 1 part Hypnorm®, 2 parts sterile water for injection, and 1 part midazolam (5 mg/mL).

[c] Prepare a cocktail by combining 1.2 mL (120 mg) of ketamine, 1.6 mL (0.8 mg) of dexmedetomidine, and 1.2 mL of sterile saline for injection. Administer at 0.25 mL/100 g BW.

[d] Prepare a cocktail by combining 0.75 mL (75 mg) of ketamine, 0.5 mL (0.5 mg) of medetomidine, and 0.75 mL sterile water for injection. Administer 0.2 mL/100 g BW. The duration of surgical anesthesia of this cocktail is reported to be directly proportional to the dose of medetomidine.

[e] The medetomidine in the fentanyl + medetomidine (neuroleptanalgesia combination) cocktail acts to lengthen anesthetic duration.

[f] Because of its toxic and carcinogenic properties, urethane solutions requires safe handling precautions (e.g., chemical fume hood) and are suitable only for nonsurvival procedures. In rats, urethane also has variably shorter duration of action, and provides variable anesthesia quality.

* Following single dose administration.

TABLE 22: INJECTABLE AGENTS SUITABLE FOR LIGHT ANESTHESIA

Anesthesia Time (min)	Agent(s)	Dose	Sleep Time (min)*
15	Pentobarbital[a]	40–50 mg/kg IV	120–240
	Tiletamine/ Zolazepam (Telazol®)	20–40 mg/kg IP	60–120
20–30	Ketamine + Acepromazine[b]	75 mg/kg + 2.5 mg/kg IP	—
	Ketamine + Diazepam	40–80 mg/kg + 5–10 mg/kg IP	120
	Ketamine + Midazolam[c]	75 mg/kg + 5 mg/kg IP	120
60	Chloral hydrate[d]	200–300 mg/kg IP of 10% solution	120–180
≥480	α-chloralose[e]	55–65 mg/kg IP	480–600

[a] Although pentobarbital is used by some investigators for anesthesia, there are more reliable agents available that have a wider safety margin and are less affected by factors such as strain and time since the last meal. For this reason pentobarbital is recommended only for light anesthesia.

[b] Prepare a cocktail with 0.75 mL (75 mg) ketamine, 0.25 mL (2.5 mg) acepromazine, and 1 mL sterile water for injection. Administer 0.2 mL/100 g BW.

[c] One may prepare a cocktail by combining 0.75 mL (75 mg) ketamine, 1 mL (5 mg) midazolam, and 0.25 mL sterile water for injection. Administer 0.2 mL/100 g BW.

[d] The use of chloral hydrate may result in peritonitis and adynamic ileus in rats.

[e] α-chloralose is not suitable for survival procedures. It has very poor analgesic properties, is reduced to trichloral ethanol (similar to chloral hydrate), and can predispose rats to seizures.

* Following single dose administration.

Note: Atipamezole administered to hasten anesthetic recovery from ketamine-medetomidine anesthesia may interfere with the postoperative analgesic effect of butorphanol.

Inhalant anesthesia

Inhalation anesthesia is very effective in rats, providing greater anesthesia control, oxygen, and shorter recovery times than most injectable agents, thus positively impacting patient survival. However, concerns include adequate waste gas scavenging, the anesthetist's skill, and equipment availability. Several inhalation agents are available, and the equipment may range from a bell jar to multiple animal anesthesia stations permitting one to anesthetize multiple rats at a time. Minimum alveolar concentration (MAC), an indication of an agent's potency, is the alveolar concentration necessary to block painful stimuli in 50% of the animals in a group. The lower the

TABLE 23: INJECTABLE ANESTHETIC ANTAGONISTS[a]

Antagonist for	Agent	Dose
Hypnorm®	Naloxone hydrochloride	0.01–01 mg/kg IP, IV
	Buprenorphine	0.01 mg/kg IV
Benzodiazepines (diazepam, midazolam)	Flumazenil	0.1–10 mg/kg IP
Opioids (fentanyl, morphine, meperidine)	Naloxone hydrochloride	0.01–01 mg/kg IP, IV
	Butorphanol	2 mg/kg SQ
	Nalbuphine[b]	2 mg/kg SQ
	Levallorphan tartrate	0.89 mg/kg SQ
Medetomidine, Xylazine	Yohimbine[c]	2.1 mg/kg SQ, IP
	Atipamezole[d]	1 mg/kg SQ, IP

[a] If reversal agents are used, both the anesthetic and analgesic properties of the drug may be terminated, thus alternative sources of analgesia must be provided.
[b] One may dilute 0.2 mL (2 mg) nalbuphine in 0.8 mL of sterile water for injection and administer 0.1 mL/100 g BW to reverse fentanyl.
[c] Yohimbine is not as specific as atipamezole. Atipamezole is preferred by the authors.
[d] Dilute 0.2 mL (1 mg) atipamezole in 0.8 mL of sterile water for injection and administer 0.1 mL/100 g.

MAC, the more potent the anesthetic agent. It should be understood that 1 MAC does not assure adequate anesthesia; however, 1.2–1.3 MAC are commonly used if no preanesthetic analgesics or tranquilizers have been given. **Table 24** provides various gas anesthetic characteristics.

- Although sevoflurane (2–2.4%) provides very rapid rodent induction and recovery, its use has been hampered by the discovery of the nephrotoxic "Compound A" (fluoromethyl-2,2-difluoro-1-trifluoromethyl vinyl ether), a byproduct of the interaction between sevoflurane and soda lime. Higher gas flows are recommended (2 L/min) to minimize Compound A production.

- Desflurane and the specialized heated vaporizer required for its use are currently expensive, thus limiting its use.

Because of the pungent odor associated with some inhalation agents, anesthetic induction is best accomplished in a suitable induction chamber such as commercially available anesthetic chambers, bell jars, or some other methods such as the one described by Gwynne and Wallace (1992). Maintenance is by either endotracheal (ET) intubation or more frequently a nose cone or facemask (commercially available, or made of a syringe case cut from a 20-mL syringe that is then fitted with a mask diaphragm made from a piece

TABLE 24: CHARACTERISTICS OF VARIOUS INHALANT ANESTHETIC AGENTS

Agent	MAC (%)	Vapor Pressure (mm Hg, 20°C)	% Saturation at 22°C	Soda Lime Stability	% Recovered as a Metabolite
Desflurane	7.2	669	92	Stable	<0.02%
Enflurane[a]	2.21	172	23	Stable	24
Halothane[c]	0.95	242	32	Slight decomposition	20–25
Isoflurane[d]	1.38	240	32	Stable	0.17
Methoxyflurane[e]	0.22	23	3	Slight decomposition	50
Nitrous oxide[f]	250	39,500	100	Stable	0.004
Sevoflurane	2.68	157	21	Some decomposition	<5%

[a] Enflurane has a lower safety margin in rats compared with other agents such as isoflurane; therefore, its use in rodents is not recommended.

[b] Using ether is not recommended because it is extremely flammable/explosive and requires specialized animal storage.

[c] Halothane has been shown to have extensive effects on the cardiopulmonary system during anesthesia.

[d] Isoflurane is very safe in rats even for long periods of anesthesia and is thus a preferred inhalant anesthetic. MAC of isoflurane decreases over 1–4 hr of anesthesia in 7-d-old but not in 60-d-old rats.

[b,e] Methoxyflurane and halothane are no longer available from vendors in the United States.

[f] Nitrous oxide is most often used as a supplemental inhalant gas anesthetic because it has insufficient anesthetic properties for surgical anesthesia and is a human health hazard.

of a rebreathing bag or glove with an appropriately sized hole). Induction doses (e.g., isoflurane 4–5%) are often significantly higher than maintenance doses (e.g., isoflurane 1–3%).

The bell jar technique ("open-drop") requires minimal skill and permits minimal control over anesthetic administration. With this, however, there is also the danger of personnel being exposed to high-level, anesthetic agents. Thus, the technique is to be discouraged, unless adequate scavenging (e.g., a chemical fume hood) is available. This technique provides only very short-term anesthesia for quick procedures such as blood collection or anesthetic induction. It involves placing the animal in an environment ("jar") where the volatile anesthetic agent evaporates. Use a container (with a secure lid) constructed of a sanitizable, nonporous material permitting constant visualization of the animal. The jar should be of sufficient size to comfortably accommodate the animal but not so large as to require excessive amounts of anesthetic. Rogers et al. (2003) described a modification of the technique using an anesthetic-soaked cotton ball in a fold-over disposable sandwich bag to anesthetize mice. The bell jar technique requires separation between liquid-phase anesthetic agent and the rat (e.g., elevated wire-grid platform or pathology cassette); the resulting contact can cause skin irritation and potential overdosing from cutaneous absorption. Methoxyflurane may be the preferred agent when using the technique, but the agent has limited availability and significant negative side effects. Meanwhile, lethal halothane and isoflurane concentrations develop under room temperature and pressure. After removal from the bell jar, one may administer additional anesthetic during a procedure by placing the shaft of a 35- or 60-cc syringe containing anesthetic-soaked cotton or gauze over the rat's face. The rat will inhale supplemental anesthetic as it breathes.

ET intubation is difficult in rodents owing to their small size and the inability to open their mouth wide enough to visualize the glottis. However, it provides an effective means to deliver adequate inhalant anesthesia for a longer duration procedure and is needed for certain procedures such as cardiothoracic surgery. The trachea can be accessed via tracheostomy or (preferably) direct passage of a tube through the mouth and into the glottis. Various nonsurgical intubation techniques are described for the rat (see Weksler et al. 1994, Samsamshariat and Movahed 1994, Linden et al. 2000, Stark et al. 1981, and Cheong et al. 2010), and a small animal intubation kit using fiber optic stylet is commercially available (Kent Scientific Corporation, Torrington, CT, USA). Direct passage of a tube can be performed blindly, but for the inexperienced operator an otoscope to visualize the larynx or a penlight placed over the external surface of the neck can be used to directly visualize the trachea. A 14 to 20-g IV catheter can serve as an ET tube, and 0.1 mL of 2% lidocaine may be placed into the tube and blown to spray onto the glottis to prevent laryngospasm. With the rat in dorsal recumbency, visualize the larynx by extending the head and retracting the tongue. Use the otoscope, place a flexible (yet firm) (e.g., a 0.038 guide wire) in the trachea, and carefully remove the otoscope. Finally, thread the catheter over the guide wire until it is in position. While holding the catheter in place, remove the guide wire and attach the catheter to the anesthetic circuit. It is very important to minimize equipment dead space in the circuit by placing the connector close to the mouth, since dead space is the compartment of gas that does not undergo physiological exchange, and if great enough, can result in the animal's suffocation.

Anesthetic and Supportive Equipment

One should be familiar with any equipment to be used to prevent anesthetic complications, hasten anesthetic recovery, and increase procedural survival rate.

Gas anesthesia machines: **Figure 55** illustrates the basic rodent anesthetic apparatus. Other setups are more elaborate and have additional features. Also, rodent ventilators can be connected to the anesthesia machine and an evacuation system, if needed. Several commercial ventilators suitable for rats are available. Respiratory rates of 60–100 breaths/min and tidal volumes of 1.5 mL/100 g should provide adequate ventilation under pressures of 5–15 mmHg.

- Monitoring equipment: During anesthesia, the patient's oxygenation, ventilation, circulation, and temperature should be continually evaluated and periodically recorded in an anesthetic record. Adequate anesthesia monitoring can indicate problems and enable corrective actions to be taken prior to the animal's death or serious impairment. Monitoring may include measurement of body temperature, respiratory rate, heart rate, arterial blood pressure, and tissue oxygenation by pulse oximetry. Although the monitoring requirements are procedure dependent, body temperature is an important parameter, because the rat's large surface-area-to-body-weight ratio predisposes it to hypothermia during surgery.

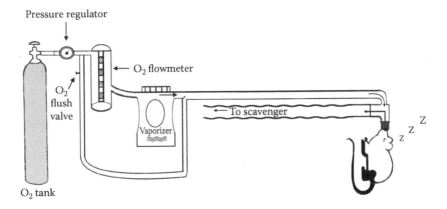

Fig. 55 Diagrammatic representation of the Mapleson E anesthetic circuit commonly used to anesthetize rodents, including rats. (Image courtesy Janis Atuk-Jones.)

• Equipment for providing supplemental heat: Avoid using electric heating blankets and heat lamps (including incandescent lamps), as they could cause burns. Roehl et al. (2011) highlight the disadvantages of electric heating pads (e.g., overheating, uneven heating), despite following the manufacturer's suggested use, and the resulting negative research implications. Strong preference is given to warm-water-recirculating blankets. Since it is much easier to keep a patient warm than rewarm it, supplemental heat is expected to be available while the animal is anesthetized. Room temperature can also be elevated (80–85°F), and when using inhalation anesthesia, tanked oxygen kept in the warm room overnight or humidification can assist in maintaining normothermia. Drapes and bubble wrap provide additional insulation to reduce heat loss. Holes can be cut in the bubble wrap, permitting adequate procedural exposure.

Neuromuscular Blocking Agents

Neuromuscular blocking (NMB) agents produce muscle paralysis, eliminating spontaneous breathing and thus requiring mechanical ventilation, so that surgery, especially intrathoracic and stereotaxic procedures can be conducted with fewer complications. NMB agents fall into two categories: depolarizing and nondepolarizing. Depolarizing agents act similar to acetylcholine and cause muscle contraction. Animals develop generalized muscular fasciculations prior to general muscle paralysis. Nondepolarizing agents compete with acetylcholine at the neuromuscular junction for receptor sites. An increase in acetylcholine reverses the blockade. Neostigmine increases acetylcholine concentration by altering acetylcholinesterase activity. Two concerns with NMB agent use are to first ensure good general anesthesia, as animals can still perceive pain, and then to provide ventilatory support, since respiratory muscle paralysis occurs. IACUCs frequently have NMB agent guidelines, and their use frequently requires scientific justification. **Table 25** lists commonly used NMB agents in rats.

Anesthesia Monitoring

Anesthesia alone and in conjunction with surgery can compromise the patient's homeostasis. Thus, careful monitoring is warranted.

TABLE 25: COMMONLY USED RAT NMB AGENTS

Agent*	Dose
Gallamine	1 mg/kg IV
Pancuronium	2 mg/kg IV
Tubocurarine	0.4 mg/kg IV

* All nondepolarizing NMB agents.

- Determine rat anesthetic depth by using the toe pinch (pedal withdrawal reflex), eye blink (palpebral reflex), skin pinch, breathing rate and depth, or a combination of these signs. Presence of reflexes may indicate inadequate anesthetic depth. Lack of withdrawal following a toe or tail pinch is the most reliable indicator of appropriate anesthetic depth. Breathing should be deep and regular.

- Intraoperative monitoring/records: Monitoring includes the physiological parameters mentioned above. When using NMB agents, heart rate should be monitored, as increases could indicate inadequate anesthetic depth. Intraoperative monitoring records coupled with postanesthesia monitoring records facilitate communication between the IACUC, animal care staff, and researchers. Postoperative monitoring records may be kept together with the cage card. These records will greatly assist with troubleshooting problems that may occur.

- Even when using sophisticated equipment, the role of basic clinical observations, such as the color of the mucous membranes and blood in the operative field and breathing depth and frequency, cannot be overestimated.

- The importance of monitoring mean arterial blood pressure and intracranial pressure in experiments on the CNS is fundamental.

- Special attention should be paid to controlling the temperature and maintaining the fluid/electrolyte balance.

aseptic surgery

General

The surgical procedure can be divided into three important stages: preoperative, intraoperative, and postoperative. One should consider each part individually when attempting any surgical procedure, identifying potential issues that could jeopardize the study or the animal's life. It is better to identify such issues in the experimental planning stages rather than after an undesirable or questionable experimental outcome. The issues could include anesthetic choice, equipment and staffing requirements, animal health status, animal-monitoring requirements, and investigator education. For novel and complicated studies, a pilot study to identify specific issues and their solutions is advisable. Veterinarians and veterinary technicians are a valuable resource in experimental planning to identify anesthetic issues. Several texts are available for in-depth discussion of aseptic technique and other surgical procedure elements (animal, surgical team, and instrument preparation; suture selection, knot tying, etc.) and these include the discussion by Cooper et al. (2000).

Surgical procedures may be classified as follows:

- Major: A procedure that penetrates and exposes a body cavity, produces substantial impairment of physical or physiologic functions, or involves extensive tissue dissection or transection (e.g., laparotomy, thoracotomy, joint replacement, and limb amputation).
- Minor: A procedure that does not penetrate and expose a body cavity and causes little or no physical impairment (e.g., wound suturing, peripheral vessel cannulation, and percutaneous biopsy).
- Survival: A procedure in which one expects the animal to regain consciousness after anesthesia.
- Nonsurvival: Animal euthanasia occurs immediately following the procedure and the animal does not regain consciousness.

Aseptic Technique

Although there is a general misconception that rodents are less susceptible to postoperative infections than other animals, aseptic technique is essential for a positive surgical outcome and to prevent pain and distress associated with postprocedural infections. General aseptic technique principles used to reduce microbial contamination to the lowest possible practical level should be followed for all survival surgical procedures. For nonsurvival surgery, the surgical site should at least be clipped, the surgeon should wear gloves, and the instruments and surrounding area be clean. Throughout this section, procedures involving aseptic technique are discussed; these involve animal, instrument, and surgical-area preparation; PPE; and the use of appropriate aseptic techniques.

Suture material

Suture material can be broadly classified into *absorbable* and *nonabsorbable*. Absorbable materials lose their tensile strength and break down by processes such as hydrolysis and proteolytic enzyme degradation. Nonabsorbable materials are made from compounds that are not metabolized by the body, and are therefore used either on skin wound closure, where the sutures can be removed after a few weeks, or in some inner tissues where absorbable sutures are not adequate.

Further suture material subclassifications include natural vs. synthetic and monofilament vs. multifilament (braided). Natural materials include catgut, steel, and silk; synthetic materials include polyglycolic acid, polydioxanone, and polyester. Preference is given to monofilament materials because multifilaments tend to draw (or wick) body fluids and bacteria into the surgical wound. In addition, multifilaments can harbor organisms among their strands.

Suture needles

Needles are available in a variety of sizes, shapes, and diameters. Typically suture needles are either tapered or cutting, with the suture material attached (swaged) to the needle. Suture material without an attached needle is available. Tapered needles are usually for internal closures, and cutting needles are for skin closure and other more dense tissues such as tendons.

Suture patterns and knots

The three most commonly used patterns are the simple interrupted, continuous, and horizontal mattress. When finishing a given pattern, the suture material is tied into knot(s), depending on a given material's knot security. One may tie these knots by hand or with instruments (needle holders) using either a square knot or a surgeon's knot, a variation of the square knot.

Suture recommendations

It is imperative that one consults a veterinarian or veterinary technician if unsure on which suture materials and patterns to use. Generally, synthetic and wire sutures are much less reactive than natural sutures. Catgut is not recommended for internal wound closure because of the potential for rapid suture breakdown following bacterial contamination. Silk is not recommended as it generates a significant tissue reaction over synthetic alternatives. Synthetic monofilaments are highly recommended for internal and external wound closure. Tissue adhesives, surgical staples, and ligating clips are available to assist in wound closure and hemostasis. One usually closes the skin with 3-0 or 4-0 suture material either with simple interrupted (preferred) or continuous suture pattern.

Fluid Therapy/Emergency Response

- Fluid and electrolyte therapy: One should administer warm fluids during a long procedure or at the end of a short procedure to ensure normothermia and/or to correct blood loss during surgery. One may administer warm lactated ringers solution (LRS) or 0.9% (normal) saline at a rate of 40–80 mL/kg/24hr SQ, IP, or IV. Considerations to any preexisting or developing condition related to the procedure (e.g., kidney failure, diabetes mellitus) should be taken.
- Cardiovascular: The adult rat blood volume (BV) is approximately 5–7% of its body weight (BW). A 25–30% BV loss (or approximately 2% of the BW) could result in shock, while a 30–40% BV loss may cause death. When using inbred strains, there is no need to crossmatch animals of the same strain prior to administering blood. One may store blood in acid citrate dextrose (ACD) at 4°C for several days prior to administration; the proper ratio of ACD to blood is 1:3. The reported successful storage time of whole blood in citrate, citrate-dextrose, and citrate-phosphate-dextrose-adenine (CPDA-1) is 8, 22, and 35 days, respectively (Wiersma and Kastelijn 1996).
- Oxygen and mechanical ventilation are needed in apnea or hypoxia to assist or replace spontaneous breathing.
- Emergency drugs: See **Table 26**.

---

preoperative issues

Anesthetic and Supportive Equipment

Ensure equipment (see Anesthesia section) is available and in good working order.

TABLE 26: SELECT EMERGENCY DRUGS

Drug	Indication	Dose
Atropine	Bradycardia, and to decrease airway secretions	0.05 mg/kg IV
Doxapram	Respiratory depression	5–10 mg/kg
Epinephrine	Cardiac resuscitation	0.1 mL/kg (1:10,000)
	Bronchoconstriction (anaphylaxis)	0.2 mg/kg IV, IM, SQ
Furosemide	Pulmonary edema	5–10 mg/kg q12hr
Dexamethasone	Shock	5 mg/kg IV bolus
Lidocaine	Cardiac arrhythmias	2 mg/kg IV

Surgical Instruments

- Sterilize items based on their physical characteristics. Autoclaving (pressurized steam), ethylene oxide (EtO) exposure, cold sterilization (chemical agents), dry heat (glass bead sterilizer), or irradiation may be used. Autoclaving and EtO sterilization are most common.

- Cold sterilization and especially bead sterilizers are valuable when procedures are performed consecutively on a group of rats. Follow the manufacturer's recommendations for contact time using these methods. When using a hot bead sterilizer, cool instruments sufficiently before touching animal tissues. The glass beads also need to be replaced regularly, usually when dirty.

- Thoroughly rinse the instruments in sterile saline or water after cold sterilization.

- "Alcohol is neither a sterilant nor a high-level disinfectant." (NRC 2010).

- Sterilization indicators should be used for sterilization validation.

- Most bulk disposable items and other surgical supplies purchased presterilized have frequently undergone irradiation sterilization.

- If one is performing procedures on groups of animals, it is best to have two or three surgical packs per surgeon, to ensure an adequate supply of sterile instruments. At the minimum, surgical gloves and drapes will also need changing between animals. If the drapes are dry, it is possible for them to go with the rat to the recovery area to aid in maintaining body temperature.

Drugs

- Check all compound expiration dates used during any component of the procedure.
- Check the concentrations of the compounds and ensure there are adequate amounts of drugs available.
- Adequate anesthetic agents should be made available in the event the procedure takes longer than anticipated.
- Make sure fluid bags, electrolyte solutions, and other emergency drugs as previously described are available.
- If using an injectable anesthetic, the appropriate antagonist, if applicable, should be available.
- Administer antibiotics prior to the procedure (prophylactic antibiotics) to ensure adequate tissue concentration at the time of the procedure. Antibiotics do not substitute for proper aseptic technique.
- Administer analgesics prior to a procedure (preemptive analgesia). This interrupts the pain cascade and reduces postoperative discomfort.

Personnel

- Adequate personnel: Enough personnel should be available to perform the procedure in both aseptic (e.g., surgeon and assistant) and nonaseptic (e.g., anesthetist, circulating technician) capacities without compromising the sterility of the procedure and integrity of the study.
- Ensure personnel have adequate information and training regarding the following:
  - Experimental protocol.
  - Anesthetic and support equipment/drugs.
  - Animals: anatomy, physiology, etc.
  - Surgical and other techniques: The personnel should be familiar with aseptic technique, perioperative management, and the surgical and other procedures in the experimental protocol.
- PPE: Optimally, the surgeon and assistant should wear shoe covers, cap, mask, sterile gloves, and gown; other individuals in the room not performing or assisting in the actual

procedure should at least wear scrubs (not exposed street clothes), shoe covers, cap, and mask. The surgeon and/or the assistant can now isolate the surgical site with sterile drapes and start the procedure using sterile instruments.

Animals

- Use healthy animals free of clinical and subclinical disease (e.g., SPF animals) to reduce morbidity and mortality, research variability, and wasted resources.

- Perform a brief physical exam to determine the animal's clinical condition.

- Permit the animals to acclimate to their new surroundings.

- One should fast rats only if performing gastrointestinal surgery (up to 12 hours); for other procedures avoid fasting longer than 2 hours.

- Animal preparation up to and including anesthesia induction and fur clipping must occur outside of the area designated for operative procedures to minimize contaminating the surgical area with bedding, fur clippings, dander, urine, and feces. One should prepare an area approximately twice the operative field needed. Once the fur is clipped, transport the animal to the surgical area.

- Once anesthetized, application of ophthalmic ointment on the rats' eyes is necessary to prevent corneal desiccation. Rat eyes are exophthalmic, which increases the risk of injury from trauma and drying during anesthesia.

Note: Withhold feed only if necessary (e.g., gastrointestinal surgery); but do not water-restrict rats. There is no need to fast rats as the rat esophagus enters the lesser curvature of the stomach through a fold in the limiting ridge of the stomach, which prevents vomiting.

---

operative issues

- This stage starts when the animal is transferred to the surgical area and positioned for final disinfection of the operative field and draping. Apply three alternating preparations with an appropriate surgical scrub (e.g., hexachlorophene, chlorhexidine, or iodophors), each followed with an alcohol or sterile water rinse. Following the last rinse, a solution containing iodine or chlorhexidine is applied to the skin 3 to 5 minutes prior to starting the procedure.

- Anesthesia monitoring (as discussed in the Anesthesia section) is important. The main goal remains to maintain a proper anesthetic depth to facilitate procedure without complications, including but not limited to death.
- Aseptic technique is essential for survival surgery.
- Gentle tissue handling and minimal tissue dissection are good techniques.
- Maintain homeostasis and normothermia: Hypothermia tends to increase anesthetic depth and sleep time.
- Minimize desiccation by keeping tissues moist with normal sterile saline.
- Before skin closure, consider administering local anesthetics (lidocaine, bupivacaine) as a line block along the incision to further reduce postoperative pain.
- Use the appropriate suture material and suture patterns for wound closure.

postoperative issues

- Monitor parameters regularly for return to normal levels: respiratory, cardiovascular, and temperature.
- Reverse injectable anesthetic agents. Postoperative administration of the injectable antagonist will minimize the anesthetic and sleep time for the procedure. Using a reversal agent (e.g., atipamezole) will also reverse any analgesia an analgesic agent (e.g., dexmedetomidine) provides, necessitating another analgesic be administered and "working" before the patient emerges from anesthesia.
- Give analgesics and antibiotics immediately after the procedure if not given preoperatively. Monitor analgesic and antibiotic needs by observing clinical signs and animal behavior and the incision site. Other surgical complications (e.g., anemia, recumbency) require regular evaluation.

- Regularly monitor wound healing by assessing incision site and checking for suture dehiscence, suture reaction, and infection.

- Support physiological needs if necessary, including providing supplementary heat and maintaining fluid balance.

- Assess surgical outcome with respect to the procedure itself and its impact upon the animal. Determine whether there are any changes that could be made to yield a better outcome.

---

euthanasia

The 2013 *AVMA Guidelines on Euthanasia* and the *Directive 2010/63/ EU* (Annex IV) provide a framework used by regulatory and accrediting bodies to describe several methods of euthanasia for animals. In 2006, the *Report of the ACLAM Task Force on Rodent Euthanasia* was published and summarizes "in a scholarly and comprehensive manner all available data-based literature relevant to these topics, to assess the scientific merit of the design and conclusions of those studies, and to compile valid information into a concise and cohesive document that could serve as a resource for diplomates, other veterinarians, IACUC members, regulatory bodies, and research scientists." Interested readers are referred to these references; more current euthanasia guidance documents or regulations may be available.

In general, euthanasia methods are categorized as

- **Acceptable:** Producing humane death when used as a sole means of euthanasia.

- **Acceptable with conditions:** Methods that by the nature of the technique or because of greater potential for operator error or safety hazards might not consistently produce humane death or are methods not well documented in the scientific literature.

- **Unacceptable:** Methods deemed inhumane under any conditions or posing substantial risk to humans applying the technique.

Laboratory rats are commonly euthanized by an overdose of inhalant anesthetics or barbiturates, or by carbon dioxide inhalation. Other commonly used conditional methods are cervical dislocation and decapitation. There are other euthanasia methods applicable to rats

TABLE 27: SUMMARY OF EUTHANASIA METHODS

Method	AVMA Guidelines	Directive 2010/63/EU (Appendix IV)
Inhalation anesthetics	AWC	Where appropriate use with prior sedation
Barbiturates	Acceptable	Where appropriate use with prior sedation
Carbon dioxide	AWC*	Gradual fill; not for fetal and neonate rodents
Cervical dislocation	AWC†	Rodents >150 g require sedation
Decapitation	AWC	Only if other methods are not possible

† Rats must weigh less than 200 g.
* Time required may be prolonged in immature and neonatal animals.
  AWC = Acceptable with conditions.

but less commonly used. Whatever method is chosen, euthanasia should be quick, with a minimum of animal distress, fear, or anxiety. **Table 27** summarizes euthanasia techniques applicable to rats.

Inhalant Anesthetics

In order of preference, isoflurane, halothane, sevoflurane, enflurane, methoxyflurane, and desflurane, with or without nitrous oxide (N_2O), are acceptable. Methoxyflurane is acceptable with conditions only if other agents or methods are not available, as it is highly soluble and causes slow anesthetic induction possibly accompanied by agitation. Personnel safety requires gas scavenging either through an appropriately equipped anesthetic machine, chemical fume hood, certified hard-ducted BSC, or ducted downdraft table. N_2O may be used with other inhalants to speed anesthesia onset, but alone it does not induce anesthesia, even at 100% concentration. Ether is not acceptable for euthanasia. The authors strongly discourage diethyl ether because of its storage requirements and the explosion hazard associated with refrigerators and freezers storing ether-exposed animal carcasses.

Barbiturates

If rats are acclimated to restraint for tail vein injection and the personnel are experienced, or the rats have patent IV catheters, euthanasia by IV barbiturates is preferred. However, if this technique is likely to cause more distress to the animal, intraperitoneal injection is acceptable, provided the solution does not contain neuromuscular blocking agents; pain may be associated with injections given the IP route. Intracardiac injection is acceptable in unconscious animals. There are several barbiturate products formulated specifically for euthanasia. In the United States, all barbiturates are regulated drugs and their use must comply with the regulations of the U.S. Drug Enforcement Agency (DEA).

Carbon Dioxide

CO_2 is commonly used because it is inexpensive and it poses minimal personnel hazard. Compressed CO_2 gas cylinders are required: CO_2 generated by other methods such as from dry ice, fire extinguishers, or chemical means (e.g., antacids) are unacceptable. Cylinders must be equipped with a regulator and properly secured to meet safety requirements. A flow meter added to the system helps determine the appropriate pressure range to assist in delivering a suitable CO_2 volume.

High concentration CO_2 inhalation may be distressing to animals because the gas dissolves in the moist nasal mucosa, producing carbonic acid, which stimulates nasal mucosa nociceptors. Displace 10–30% of the chamber volume per minute to provide an optimal flow rate. Prefilled chambers are not recommended. Since CO_2 is heavier than air, incomplete chamber filling causes animals to climb or raise their heads to avoid CO_2 exposure. It is recommended to wait for at least one minute following respiratory arrest to assure the animal is dead prior to use or disposal. Commercial euthanasia systems are available either as manual single-cage or automated CO_2 delivery systems that can fit multiple cages. Chambers must not be overcrowded, and avoid mixing unfamiliar animals. Rats with respiratory disease and neonatal animals are resistant to the effects of CO_2.

Cervical Dislocation

This causes direct brain depression, may induce rapid unconsciousness, and does not chemically contaminate tissues. Rats must be less than 200 g (AVMA) or 150 g without prior sedation (Directive 2010/63/EU). The rat is grasped behind the skull or a rod is placed at the base of the skull and then the base of the tail or hind legs are pulled. This technique must be performed by skilled personnel. Personnel should be trained on anesthetized and/or dead animals to demonstrate proficiency.

Decapitation

Decapitation causes direct depression of the brain and leaves it intact. Guillotines are commercially available. They should be maintained sharp and in good repair and be used by well-trained personnel. The rat must be restrained (e.g., in a disposable plastic cone) such that there is little distress to the rat and little risk to the handler from either a rat bite or a guillotined finger. Drop the blade quickly and firmly to ensure a quick cut, not a slow crush. Rats are sensitive to the smell of blood, and one must clean the guillotine well before proceeding to the next animal. If not, the technician may find subsequent rats more aggressive and difficult to restrain. Rats remain calmer if brought into the room individually rather than as a group.

Parenteau et al. (2009) evaluated the effects of short isoflurane anesthesia immediately prior to decapitation on CNS penetration/ distribution of reference and new compounds and determined that isoflurane has no effect. Sprague (1993) describes a centrifugation-induced analgesia technique that can be used immediately prior to decapitation.

Microwave Irradiation

Microwave irradiation is an acceptable rat euthanasia method used primarily by neurobiologists to fix brain metabolites *in vivo* while maintaining the brain's anatomic integrity. Appropriate microwave units, and not those designed for kitchens, must be used. All units direct their microwave energy to the animal's head. Unconsciousness is achieved in less than 0.1 second, and death in less than 1 second.

Fetal and Neonatal Euthanasia

The following are acceptable euthanasia methods:

- Fetuses up to 15 days of gestation: Euthanasia of the mother or removal of the fetuses causes fetal death from loss of blood supply or nonviability of fetuses at this stage of development.
- Fetuses over 15 days of gestation:
  - Injection of chemical anesthetics; decapitation with sharp surgical scissors; cervical dislocation; chemical fixation of anesthetized fetus.

- When fetuses are not required, the dam may be euthanized using an adult method that ensures cerebral anoxia to the fetuses and minimally disturbs the uterine milieu to minimize fetal arousal (e.g., CO_2 exposure followed by cervical dislocation).
- Up to 10 days of age: Anesthetic injection, decapitation, cervical dislocation, immersion in liquid nitrogen preceded by anesthesia, or perfusion with chemical fixatives under deep anesthesia. CO_2 may require exposures as long as 50 min; therefore, consider using another method.
- Over 10 days of age: Follow adults guidelines.

Nonsurvival Surgery

There are instances (e.g., microsurgery training classes) where anesthetized animals are euthanized after surgery and before anesthetic recovery. When conducting nonsurvival surgery, at a minimum the surgical site should be clipped, the surgeon should wear gloves, and the instruments and surrounding area should be clean. For long surgical procedures, aseptic technique may be important to ensure model stability and successful outcome. Additionally, animals may be perfused with a fixative (e.g., glutaraldehyde); this procedure is performed under deep anesthesia (e.g., pentobarbital 80–120 mg/kg IP) and will result in the animal's death. The perfusion area must have appropriate safety equipment and undergo regular inspections by environmental health and safety (EHS) and IACUC personnel. The safety equipment depends on the various and cumulative animal hazard exposure encountered during the research paradigm, including perfusion. EHS will provide guidance on proper handling and disposal of the fixative material and/or the animals' remains.

necropsy

Procedure

In general, conduct the necropsy soon after the animal's death and before postmortem changes (PMC) occur. PMC include putrefaction and autolysis, and either one may predominate, depending on the circumstances surrounding death and the environmental conditions. Putrefaction involves the bacterial action on body tissues and is associated with the body's green discoloration, gas production with associated bloating, skin slippage, and foul odor. Autolysis, the breakdown of the body by endogenous substances, proceeds most rapidly in organs such as the pancreas and stomach.

If an infectious disease is suspected, the preferred method is to humanely euthanize a rat with clinical signs. This allows the prosector to take diagnostic samples (e.g., blood for serology) and ensures that bacterial pathogens are not overgrown by those bacteria associated with putrefaction. It is highly recommended to consult a veterinarian before necropsy, especially if infection is suspected. For toxicological studies it is imperative to avoid any PMC, which could compromise histologic interpretation. Carcass refrigeration will slow PMC, not prevent it. Freezing the carcass is generally unacceptable because of the resulting tissue architecture destruction ("freeze artefact"). Detailed descriptions of necropsy techniques are available elsewhere, including the guide by Feldman and Seeley (1988). The following is a brief outline of a rat necropsy:

1. Proper necropsy reports or data sheets should be used to record necropsy findings (see below). Clinical and experimental history should be obtained, as these might be informative about which organs to collect and for later interpretation of necropsy and histopathologic findings.

2. The rat is weighed, and any animal identification and/or protocol number is recorded.

3. A pre-necropsy exam is performed, consisting of a visual inspection of the pelt, limbs and extremities, all orifices, and palpation of the body. Note any abnormalities, including superficial swellings, masses, injuries, wounds, and skeletal deformities.

4. Wet the ventral fur with alcohol to facilitate the dissection. At this point, cardiac puncture can be performed for blood collection when the animal has been euthanized as opposed to being found dead. A quick method to view the organs *in situ* is to take surgical scissors and cut through the skin and muscle of the lower abdomen. Extend the incision on both sides of the abdomen toward the thorax and through the ribs. This exposes all major thoracic and abdominal viscera. Further extension of the incision provides visualization of the ventral cervical region and such associated organs as the salivary glands.

5. Visceral cavities are examined for any abnormalities, including pus, blood, effusion, and other excess fluid (e.g., urine), and if any are present, it is advisable to quantify and sample them for diagnostic analysis. Visceral organs are examined, noting their sizes, shapes, textures, and colors. Give special attention to the teeth, salivary glands, Harderian glands, lungs, intestinal tract, and genitourinary tract, as these are common tissues affected by rat diseases.

6. Immersion-fix routine histopathologic samples in 10% neutral buffered formalin at a ratio of 1 part tissue to 10 parts formalin. Inflate the lungs with fixative to preserve the state of pulmonary expansion and the structural relationships within the lung parenchyma. The intestinal tract is also filled with fixative for adequate fixation. Choice of fixative varies depending on the purpose of the necropsy; for example, a better fixative choice for testes evaluation may be Bouin's solution. Areas with bone can be placed in commercially available solutions combining fixation and decalcification (e.g., Formical-4™). Alternatively, the brain and spinal cord can be removed intact and fixed by setting the rat set against a rigid support mat. Regardless of fixative, samples should be small, less than 1 cm thick, to allow adequate fixative penetration. Optimal time for formalin fixation for most histologic stains is 3–7 days. After fixation, tissue can be stored in 70% ethanol. The prosector must check with the pathologist or researcher evaluating the tissues to determine the best preservation method. Brain studies may require whole-animal perfusion with formalin or other suitable fixative.

Documentation

Each institution should develop a method for documenting necropsy findings (**Figure 56**), including appropriate submission of documents comprising a complete clinical and experimental history. Electronic record databases facilitate providing the prosector/pathologist the clinical and experimental background (Sharp and Reed 2001 2002, Sharp et al. 2005). Documentation must include records of any samples (gross or histopathologic) or photographs taken for further evaluation; such information will be compiled for each case and included, as needed, in a final pathology report.

Great Garden State University
Comparative Medicine
Diagnostic Laboratory Necropsy Evaluation Worksheet

Number Examined	Sex: M F	Protocol Number	Accession Number
	Weight:		— —

Animal ID	Species	Strain

Pre-Necropsy Examination
(N = Normal; — = None)

1	Hair Coat/Skin		5	Diarrhea	
2	Skeletal Palpation		6	Hydration	
3	Nasal Discharge		7	Body Fat	
4	Ocular Discharge		8	Ears	

Please enter number of any organ system abnormalities below followed with an explanation.

Gross Necropsy Examination
(NGL = No Gross Lesions; NE = Not Examined; NA = Not Applicable)

1	Respiratory System		10	Spleen	
2	Digestive System		11	Lymph Nodes	
3	Musculoskeletal System		12	Harderian Glands	
4	Urinary System		13	Adrenal Gland	
5	Genital System		14	Thyroid	
6	Heart		15	Pituitary	
7	Brain		16	Middle Ear	
8	Spinal Cord		17	Eye	
9	Thymus		18	Other:	

Please enter number of any organ system abnormalities below followed with an explanation.

Cellophane Tape Test:	Pelt Examination:
Extent of Postmortem Degeneration:	None Mild Moderate Severe

Fig. 56 A gross necropsy record organizes the pre-necropsy and gross necropsy findings by highlighting to the prosector the organ systems to be examined. It assembles information on the histopathology, serology, bacteriology, mycology, parasitology, molecular aspects, etc. of samples collected during the necropsy.

Diagnostic Testing

Tissues Taken For Histopathology

Lung		Brain		Testes		Ovaries	
Heart		Liver		Adrenal		Uterus	
Stomach		Kidney		Thyroid		G. Bladder	
Duodenum		Spleen		Lymph Nodes		Ureters	
Jejunum		Pancreas		Spinal Cord		Middle Ear	
Ileum		U. Bladder		Trachea		Nasal Passages	
Cecum		Pituitary		Salivary gland		Harderian gland	
Colon							

Serology (Please Circle)

Full Mouse:	MHV	Reo 3	EDIM	PVM	*M pulmonis*	MNV
	GD VII	LCMV	Ectro	MPV	MVM	Sendai

Full Rat:	SDA/RCV	Sendai	*M pulmonis*	RPV	LCMV	PVM
	RMV	RTV	*P. carinii*	H-1		

Additional Testing: Please list the collection site under each test

Bacteriology	Fungal	Parasitology	PCR/Molecular

_____   _____

Signature of Prosector   Date

Additional Comments:

_____
_____
_____
_____
_____
_____
_____
_____
_____

Fig. 56 (Continued)

experimental methodology

Some rat experimental techniques are fundamental. However, through the years, there has been a paradigm shift to develop and employ better and more refined techniques. Commercially available training rat dummies (**Figure 57**) (e.g., MD-PVC,® Squeekums,® and Koken rat®) can replace live animals for teaching and practicing various techniques.

restraint

Manual Restraint

Frequently handling rats in a firm, yet gentle, manner will accustom ("gentle") them to manipulation. Remove rats from their cage or primary enclosure by grasping the tail 1–2 cm from the base. Grasping the rat near the tail tip can cause a degloving injury. Once removed from the cage, the rat can then be placed on a flat surface for manual restraint.

One may restrain the rat by any of the following means (**Figure 58**):

- Firmly grasp the rat around the thorax with thumb or index finger under the mandible to prevent being bitten.
- The rat may be palmed over its back with the index and middle finger placed between its mandible to minimize the chance of being bitten.
- Wearing a laboratory/surgical gown and while grasping the tail with the dominant hand, hold and palm the rat against your body by the side of the hip by using the nondominant

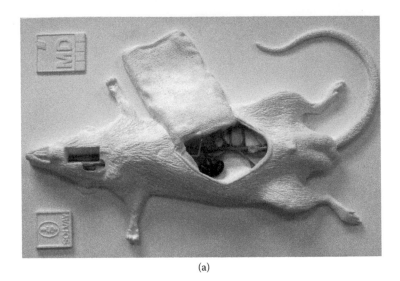

(a)

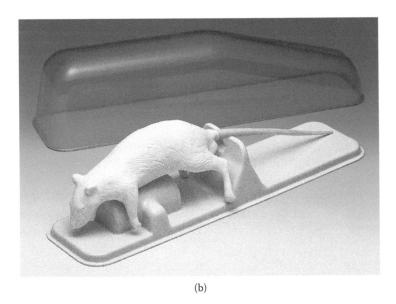

(b)

Fig. 57 Microsurgical Developments Foundation's MD-PVC Rat®
(a) offers an alternative for those using rats for microsurgery training.
The Koken Rat® (b) can be used for training in various nonsurgical
experimental techniques (e.g., gastric gavage, tail vein injection).

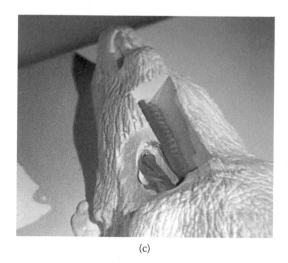

(c)

Fig. 57 (*Continued*) Microsurgical Developments Foundation's MD-PVC Rat® (c) This is a close-up of Figure 57(a). (Images courtesy Braintree Scientific, Inc.)

(a)

Fig. 58 Manual rat restraint techniques: (a) grasping the rat around the thorax with the index finger under the mandible.

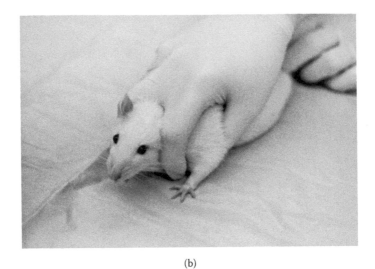

(b)

(c)

Fig. 58 (*Continued*) Manual rat restraint techniques: (b) palming the rat over the back while keeping the rat's head between the middle and index fingers; (c) holding and palming the rat against one's body for an IP injection.

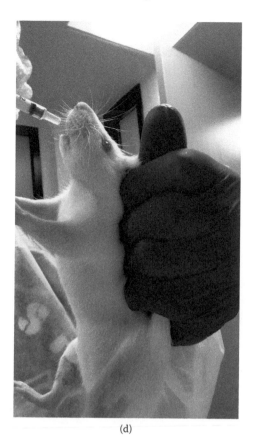

(d)

Fig. 58 (*Continued*) Manual rat restraint techniques: (d) grasping a rat by the loose skin over the back and neck.

hand. With the rat slightly facing downwards, place the index finger and the thumb caudal and cranial to the hind leg, respectively to isolate the leg. This is a good restraint technique for IP, IM, and SQ injections and saphenous blood collection.

- Grasp the loose skin over the back and neck with the non-dominant hand.

Restraint with the nondominant hand leaves the dominant hand free for animal manipulations or supporting the animal's hindquarters. With single-hand restraint, the tail may be held between the fourth and fifth fingers.

Mechanical Restraint

Several commercial mechanical restraint devices are available (**Figure 59**) and are designed for short-term handling and include various rigid plastic adjustable restrainers, decapitation cones, and drapes ("rat wrap"). These devices provide good general restraint, with the rigid restrainer and decapitation cone providing good visualization and access for injections.

Chemical Restraint

Chemical restraint, discussed in detail in Chapter 4, may prove useful for chronically painful procedures or those requiring strict

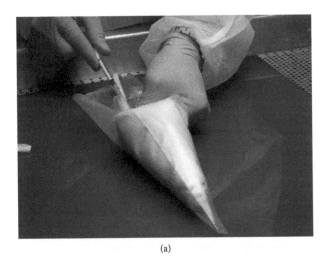

(a)

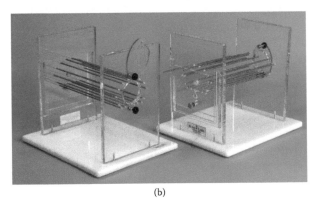

(b)

Fig. 59 Mechanical rat restraint techniques: (a) decapitation cone (b) Bollman apparatus.

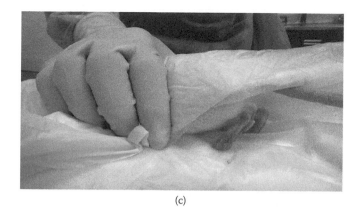

(c)

(d)

Fig. 59 (Continued) Mechanical rat restraint techniques: (c) using a drape for the "rat wrap"; (d) Broome-style plastic restraining tubes.

immobilization. As a general rule consider inhalant anesthesia, especially for short procedures.

sampling techniques

Here we present commonly used techniques for collecting various samples from rats. Interested readers are referred to the text by Waynforth and Flecknell (1992), Weiss et al. (2000), and Koch (2006) for detailed descriptions of other more invasive sampling techniques, including those used to collect bile, lymph, saliva, and pancreatic juice.

Blood Collection

General blood collection guidelines

- Blood collection sites, the instruments needed, and restraint methods are found in **Table 28**.
- The blood sampling technique is selected on the basis of anticipated animal discomfort, the blood volume (BV) needed, the blood analyses requirements, and personnel proficiency.

TABLE 28: BLOOD COLLECTION METHODOLOGY SUMMARY

Blood Collection Site	Instruments Needed	Restraint Method
Cranial vena cava	Syringe with 1" 25-ga needle	General anesthesia
Cardiac puncture*	Syringe with 19–21-ga needle	General anesthesia
Decapitation*	Guillotine or autopsy shears (neonates)	General anesthesia
Dorsal metatarsal vein	23–25-ga needle/5.5-mm lancet	Restraint device and/or manual restraint
Facial (submandibular) vein	21-ga needle/5.5–8.0-mm lancet	Manual restraint/general anesthesia
Jugular vein	Syringe with 21-ga needle	Manual restraint/general anesthesia
Retro-orbital plexus	Capillary tube/Pasteur pipettes	General anesthesia
Saphenous vein	21-ga needle/5.5–8.0-mm lancet	Restraint device and/or manual restraint
Sublingual vein	23-ga needle/5.5-mm lancet	General anesthesia
Tail vein/artery	21–25-ga needle, over-the-needle catheter, or butterfly catheter	Restraint device and/or manual restraint

* Terminal procedure only.

- An adult rat's blood volume (BV) is approximately 5–7% of its body weight (BW). Most animals will go into shock if 25–30% of their BV (or approximately 2% of the BW) is collected over a short period of time. Removing 30–40% of BV may cause death in at least 50% of animals.

- As a guide, a single survival blood collection should be limited to 1.25 mL/100 g BW. If this maximum volume is taken, then a 2-week interval between collections should be safe.

- For multiple/chronic blood draws, a total of 0.5% of the animal's BW can be removed.

- One should be cautious of physiologic effects of age, stress, and experimental manipulations.

- It is possible to evaluate whether an animal has sufficiently recovered from blood collection by monitoring the hematocrit, or packed cell volume (PCV). Note that PCV will not show alterations acutely.

- Approximately 3–4% BW (50–75% of total BV) can be obtained through exsanguination. Giving fluids during bleeding to maintain the animal's blood pressure can increase the total volume of blood cells obtained. Deep anesthesia is required prior to exsanguination.

- Perform neonatal rat blood collection by severing the jugular and carotid vessels, decapitation, amputation of an extremity, or cardiac puncture through the thoracic inlet. Collect fetal blood by decapitation or cardiac puncture after opening the thoracic cavity.

- Blood withdrawal from superficial vessels (e.g., tail vein) may be facilitated by vessel dilation via
  - increasing the temperature of a particular body area by applying a hot damp cloth or gauze
  - using a tranquilizer or local anesthetic cream
  - gentle finger-tapping of the skin overlying the vessel

- It is advisable that the blood collection site be disinfected or at least swabbed with ethyl alcohol, as the skin, the body's primary defense, is breached.

- For survival techniques, one should start distally (farthest from the heart) and move proximally in case repeat collection is necessary.

- Collect serial blood samples through survival techniques or using indwelling jugular or femoral catheters.
- The rat should not be returned to its cage until complete hemostasis has been achieved using direct digital pressure. Arterial punctures may require up to several minutes of pressure.

Note: Blood sampling and restraint techniques could affect hematology, clinical chemistry, and other test results; see Chapter 1.

Automated blood sampling (ABS)

ABS is used most frequently for drug metabolism and pharmacokinetic studies. It offers many advantages, including reduced animal stress, higher throughput, and reduced animal numbers required for a given result—all supporting the 3Rs. The equipment (**Figure 60**) permits serial collection at predetermined time points of liquid blood, dried blood spot (DBS) samples, bile, and telemetry data while the animal is tethered but unrestrained in an enclosure. Units with multiple channels that permit, among other things, simultaneous infusion and sample collection are available. Blood samples (e.g., 200 µL) may be split between liquid blood (e.g., 180 µL) and DBS (e.g., 20 µL), while storing samples at 4–6°C and room temperature, respectively. Telemetry information includes locomotion, body temperature, heart rate, blood pressure, EEG, and ECG.

Animals require previous cannulation/catheterization for sample/data collection, which can be performed either in-house or by the animal vendor (e.g., Charles River, Taconic). Furthermore, arterial catheterization appears to provide higher patency over time than venous (Mason-Bright 2011). The catheters require periodic maintenance to remain patent, including heparin lock solutions (e.g., heparinized dextrose [500 IU/mL], heparinized glycerol [500 IU/mL]), periodic flushing (e.g., every 3–5 days), and cleaning catheter exteriorization sites with dilute chlorhexidine solution followed with an antibiotic spray (e.g., gentamicin). The catheter maintenance methods and practices recommended by the vendors who performed the catheterization should be followed.

Tail vein

Accessing the lateral tail veins are the easiest to perform and best tolerated by the rat. To best visualize the vein, vasodilate it using warm water (38–40°C for 1–2 min) or with careful use of a heat lamp (1–2 min). A 21- to 25-ga needle (**Figure 61**), over-the-needle catheter,

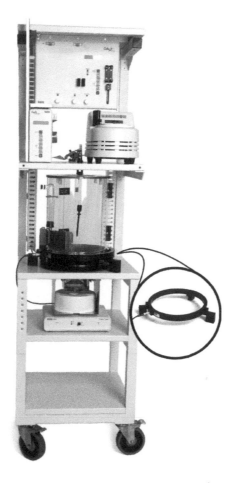

Fig. 60 Automated Blood Sampling with Telemetry unit. 495 mm ×
495–953 mm × 1778 mm.

or butterfly catheter with the tubing cut-off may be used. If needed,
a 2-mm incision can be made over the tail vein and blood collected
using a capillary tube, pipette, or a collection tube. A 1- to 2-mm
tail snip may be performed without local anesthesia for neonatal or
juvenile rats and with local anesthesia for older rats. Bleeding from
the sample site must be controlled using electric or chemical (styptic
powder or silver nitrate) cautery methods.

Jugular vein

The anesthetized animal is placed in dorsal recumbency, with the
area over the jugular clipped and prepared. An incision is made
over the jugular vein just as it passes the pectoral muscle. Following

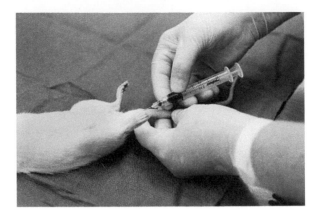

Fig. 61 Tail vein blood collection.

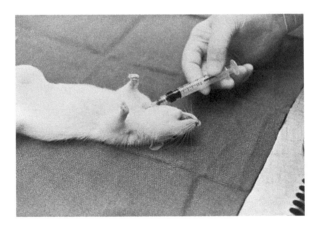

Fig. 62 Jugular vein blood collection.

removal of the sample, the site is closed with suture, wound clips, or tissue adhesive. Alternatively, when performed without anesthesia the hair over the area is clipped (usually the right side), the neck is extended, and the skin (on the right side) is entered at an acute angle with a 21-ga needle (**Figure 62**). This technique may require practice proficiency and may require anesthesia for novices.

Retro-orbital plexus

Although retro-orbital blood collection is quickly mastered, it has largely been replaced by other methods for humane reasons. With the nondominant thumb and index finger used to hold an anesthetized

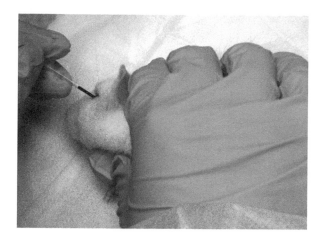

Fig. 63 Retro-orbital blood collection.

rat's head steady, apply gentle caudal traction to the eye to protrude it while the thumb occludes the jugular vein. A capillary tube or pipette is positioned at the eye's medial canthus and the ocular conjunctiva and the underlying orbital plexus is entered caudomedially (**Figure 63**). Should blood fail to flow, the collector is either not in the plexus or is applying too much pressure, preventing blood from collecting in the tube or pipette. When the volume is collected, remove the tube or pipette and remove the thumb's pressure on the jugular vein. It is also sound to apply a small quantity of antibiotic ophthalmic ointment to the eye to aid hemostasis and reduce the incidence of infection.

Properly performed orbital puncture results in minimal trauma to the animal and its eyes. However, it would more likely cause more tissue damage than venipuncture. Lesions from a single retro-orbital bleeding appear to heal without appreciable scar tissue within 4 weeks. In one report, rats with two retro-orbital bleeds within 2 weeks had hemorrhagic remnants and an increase in connective tissue fibers. The degree of discomfort caused by orbital puncture was reportedly similar to those by saphenous and tail vein bleeds, based on behavioral changes.

Saphenous vein

The lateral saphenous vein (**Figure 64**) runs dorsally and then laterally over the tarsal joint. Immobilize the hind leg in the extended position by applying gentle downward pressure immediately above the knee joint. This stretches the skin over the ankle and makes it

Fig. 64 Lateral saphenous blood collection.

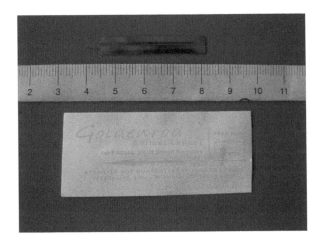

Fig. 65 Lancet used for facial vein blood collection.

easier to clip the hair over the area and immobilize the saphenous vein. The authors have also used the medial saphenous vein for rat blood collection.

Facial vein

More prominently employed for mice, submandibular bleeding can also be used for rats. Commercially available lancets (**Figure 65**) greatly facilitate the procedure (**Figure 66**). The rat is held tightly by the scruff with the whole hand and fingers. For albino rats, the land-mark is a gray dot located on the jaw line, directly below the lateral canthus of the eye. Insert the lancet or 21-ga needle at a 30° angle, 0.5 cm behind the gray dot. For pigmented rats, insert the lancet at

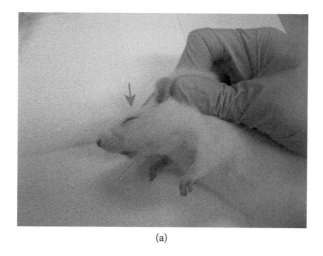

(a)

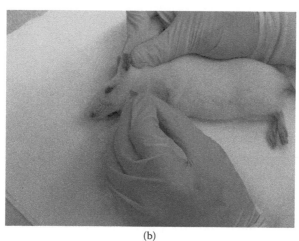

(b)

Fig. 66 Facial vein blood collection. (a) Landmarks include the lateral canthus of the eye (blue arrow) and the small circular hair tuft (red arrow). (b) Firmly scruff the rat behind the ears as demonstrated. Insert the lancet just caudal to the landmark.

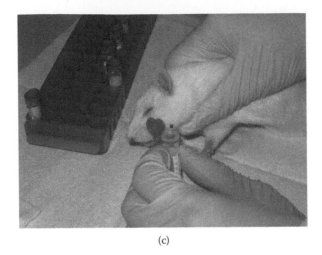

(c)

Fig. 66 (*Continued*) Facial blood collection. (c) Collect the blood in an appropriate blood collection container. (Images courtesy John Arzadon.)

a point from a line drawn straight down from the lateral canthus of the eye to the jaw line.

Dorsal metatarsal vein

This is a convenient location for small-volume blood collection and may be performed with an assistant restraining the rat in one hand and firmly extending the leg with the other. Shaving the dorsal surface of the hind foot permits better visualization, as will having the assistant use his or her thumb to distend the vessel at the level of the ankle and keep the skin tight. A hypodermic needle or lancet is inserted into the vein until blood flows (**Figure 67**).

Sublingual vein

Place the anesthetized rat in a supine position. An assistant can gather up the loose skin at the nape of the neck to produce partial stasis of the venous return from the head. Grasp and extend the tongue with thumb and forefinger. Do not use forceps, as this would cause blunt trauma to the tongue. The two sublingual veins are clearly visible at the base of the tongue and are situated left and right of the median line. Puncture a vein as far peripherally as possible, and carefully to avoid sticking the needle completely through the vein. For repeated sampling, the sublingual veins can be punctured alternately and closer to the base of the tongue. After

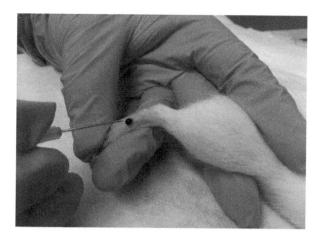

Fig. 67 Blood collection of the dorsal metatarsal using a hypodermic needle.

a successful puncture, turn the rat back into a prone position and allow the blood to drip into a tube. When the required blood volume has been collected, cease the compression by releasing the scruff of the neck.

Cranial vena cava

Jekl et al. (2005) describes blood collection through the rat cranial vena cava. Briefly, insert the needle just cranial to the first rib, 0.3–0.8 cm lateral to the manubrium at a 30° angle in the direction of the opposite femoral head. Insert the needle another 0.2–1 cm until blood begins to flow. No serious complications such as those observed in other experimental animals (e.g., vascular lacerations, heart puncture, serious hemorrhage, tracheal and throat trauma) were observed in the study.

Decapitation

One can rapidly collect large volumes of mixed venous and arterial blood from rats using decapitation, a method of euthanasia approved with conditions. Collected blood is also usually contaminated with tracheal and salivary secretions.

Cardiac puncture

The rat heart is midline in the thoracic cavity and accessible by thoracotomy, the left side of the chest, through the diaphragm (**Figure 68**), or from the top of the sternum. When approaching

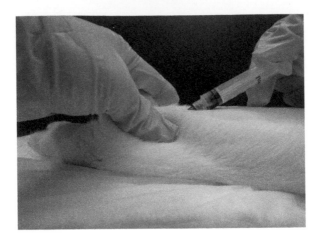

Fig. 68 Posteuthanasia cardiac puncture.

the heart from the left side of the chest, insert the needle between the 3rd and 5th ribs, or at the point of maximum heart palpitation. Use cardiac puncture only for terminal blood collection because of the risk of cardiac tamponade, pulmonary hemorrhage, and pneumothorax.

Urine Collection

Urine is typically collected through manual bladder expression, cystocentesis, vesicular catheterization, or placing the rat in a metabolism cage (**Figure 10**), stainless steel examination table, zip-lock bag, or carton box. One may easily perform vesicular catheterization in females; however, it is impossible to perform in males, because of the curves of the urethra.

Manually restraining a rat will yield a small urine sample. If this quantity is insufficient, one can apply pressure to the lower abdomen over the urinary bladder with increasing, but not excessive, force to manually express the bladder. Cystocentesis can be attempted but requires palpation and isolation of the rat's small urinary bladder. Thus, it is recommended that cystocentesis be used only at necropsy in rats. Vesicular catheterization is possible in females with a 22-ga, 5-cm, over-the-needle catheter. Perform the procedure aseptically to avoid the introduction of pathogens into the urinary bladder. The anesthetized rat is held in the nondominant hand with the head facing the investigator, and the tail looped around the nondominant index finger. The nondominant thumb applies gentle cranial tension to the lower abdomen to make the urethral opening apparent. The catheter is advanced caudally initially as the urethra rises over the pelvis, then cranially into the bladder.

Cerebrospinal Fluid

CSF may be collected from lumbar or cisternal punctures, as a survival or nonsurvival procedure, and through chronic lateral cerebral ventricle cannulation. Some procedures described require the use of a specially designed holding apparatus for the rat (Lebedev et al. 2004; Petty 1982). When CSF collection volume is less than 50 µL in F344 rats, the collected CSF is reportedly of better quality. The typical low CSF yield is no longer a limitation for biomarker measurement in toxicology and neurological disorder research because of modern molecular techniques. Sharma et al. (2006), Strake et al. (1996), Sanvitto et al. (1987); and the text by Petty (1982) are good resources for various collection techniques. Rago et al. (2011) describes CSF dried spot collection and analysis.

Bone Marrow

Bone marrow collection is done under general anesthesia using aseptic technique, including the surgical preparation of collection site. One may use a low-speed dental drill to create a small hole in either the tibia or femur large enough to insert a polyethylene tube or spatula tip. Biopsy needles (18 ga, 1") can also be used to collect samples from the iliac crest. One could also use the sternum, or the bones of the proximal one third of the tail. Further discussion on bone marrow collection may be found in Tavassoli et al. (1970), Šulla et al. (2008), and Bolliger (2004), the last providing a rat-specific overview of bone marrow evaluation and morphology.

Milk Collection

Milk collection may be necessary for studying compounds that pass in the milk, fostering pups, or dosing pups in a milk medium. The simplest of the techniques involves repetitive stroking motions over the teat and manually stripping to collect milk. Other methods employ the use of apparatuses with vacuum systems such as the ones described by Rodgers (1995) and Waynforth and Flecknell (1992).

Feces

Rats can be housed in metabolic cages or wire-bottom cages for fecal collection. However, samples are frequently contaminated with urine and fur. It is important to note that rats are coprophagic and consume feces even when housed on wire-bottom cages. Special techniques, described by Waynforth and Flecknell (1992), are then necessary to prevent coprophagy and collect all voided feces. To obtain small numbers of fecal pellets, the rat can also be manually restrained and/ or the rectum stimulated, but this technique depends on the presence of feces in the rectum at the time of collection. Individual animals may be placed briefly in a clean cage (without bedding) to collect feces.

Macrophage Collection

Freshly euthanized rats undergo one or two peritoneal lavages (e.g., 5–15 mL each) with tris-buffered Hank's solution or Hank's balanced salt solution.

Lymphatic Fluid

Lymphatic fluid collection may be of interest in intestinal absorption, pharmacokinetic studies, and immunology studies, and usually requires lymph sampling in awake and unrestrained rats. The procedure is challenging, in part because the lymphatics are delicate structures. Rats receive 0.5–1.0 mL of olive or corn oil by gastric gavage 0.5–1 hour preoperatively to stimulate lymph flow and to aid in visualizing the lymph ducts. Several of the described approaches for thoracic duct cannulation in rats have modest reproducibility and a low lymph flow rate. Ionac (2003) described an improved method for obtaining thoracic duct lymph using an operating microscope and silicone cannula and the maintaining animal mobility during lymph collection. Ionac (2003) also described the thoracic duct anatomy. Mesenteric lymph can be collected by methods requiring total rat restraint and fluid replacement by intravenous or intraduodenal infusion to maintain lymph output, or from conscious, minimally restrained animals, as described by Hauss et al. (1998).

Bile

Bile is usually collected in pharmacokinetic and metabolism studies because many test compounds are metabolized by the liver and excreted in the bile. Several bile collection methods have been described for chronic bile collection. Whatever method is used, one should note that since the rat lacks a gall bladder, the common bile duct (CBD) carries both bile and pancreatic juices. Thus, to collect pure bile, the CBD should be catheterized near the liver and before the pancreatic ducts enter the CBD. The cannula can be fixed by a tail cuff or exteriorized in the scapular region through a jacket or harness. Faure et al. (2006) described catheter implantation into the bile duct to divert the bile flow via an exteriorized loop in freely moving rats. The collected bile may be re-infused or replaced with artificial bile salt solutions to avoid bile salt depletion.

Vascular Access Port

A blood vessel is catheterized and the port is implanted subcutaneously to prevent catheter dislodgment, improve catheter patency, and reduce infection problems associated with exteriorized catheters. Once implanted properly, VAPs can be maintained in a comfortable and sterile condition for chronic studies, and animals can be group-housed since the risk of damage to external implants is eliminated. VAP use has evolved into a multipurpose access port for use in intestinal, biliary, intraspinal, cranial, ventricular, and other studies. MRI-compatible VAPs are available for survival MRI studies requiring repeated compound administration, intravenous contrast, or blood sampling. Although the paper by Swindle et al. (2005) primarily discusses VAP use in large animals, it provides a thorough discussion of basic VAP principles and technology and their use in animal research. The catheter may be pre-attached during VAP manufacture or attached intraoperatively by the surgeon. Rats with implanted VAPs can be purchased from commercial vendors as well. VAPs are available in a variety of configurations. The chamber has a thick polymer septum that can withstand repeated punctures, and has a self-sealing action when using the specially designed (Huber) needles.

VAP success depends on the device's design, proper surgical and maintenance technique, and consideration of species-specific behaviors. Disadvantages of VAPs include costs, the presence of "dead volume" within the port, the need for a meticulous aseptic implantation technique, the risk of necrosis or port erosion through the skin, and the potential for the needle to dislodge during infusion.

VAP alternatives include the Vascular Access Harness™ (Instech Laboratories, Plymouth Meeting, PA), which is a system with a miniature external port housed in a harness and a mating spring tether with a connector; and a system with a silicone catheter connected to an IV indwelling cannula with fixation wings and a closure stopper with injection port (de Wit et al. 2001).

compound administration

A compound's administration route is influenced by the purpose of the compound administered, pharmacology, and other considerations highlighted below:

- pH:
  - The pH should be 4.5–8.0.
  - The greatest pH tolerance is seen with IV administration (owing to blood's buffering capacity), followed by IM, and then SQ administration.
  - Solutions less acidic than 0.1 N HCl are well tolerated by oral administration.
  - Alkaline solutions are not well tolerated by the stomach.
  - pH is but one measure of a compounds potential irritancy. Irritant compounds/vehicles must be used with caution when administering them at various sites, including IP, intratracheal/intranasal, and the periocular region.
- Solubility and nature of the compound (including vehicle): Water-soluble compounds may be mixed with sterile water, 5% dextrose, or normal saline, whereas nonwater soluble compounds may be used with propylene glycol or DMSO (dimethylsulfoxide) as a vehicle. One may deliver water miscible vehicles by any route, whereas those vehicles that are oils should not be administered IV. Certain vehicles, such as 5% dextrose, may need reconsideration when planned for chronic continuous IV administration owing to the potential for opportunistic infections.
- Volume: **Table 29** provides guidelines for optimal compound administration volumes.

When administering injectable compounds, one should use the smallest gauge needle possible and the most accurate and appropriate syringe for delivering the necessary volume. The needle gauge will depend on the solution's thickness or viscosity, with more viscous

TABLE 29: RAT VOLUME ADMINISTRATION GUIDELINES

Route	Volume	Notes
Gastric gavage	20 mL/kg	Higher volumes, as much as 40 mL/kg BW, can cause undue stress to rats and must be administered with caution. If large volumes are to be given, it may be necessary to fast the rat before dosing. For nonaqueous liquids, the dose volume should be reduced by 50%.
Intracerebral	20 µL or 2% of brain volume	For stereotactic procedures, rates of 0.5–1 µL/min are reported with up to 5 min between injection cessation and needle withdrawal.
Intracolonic	3.5 mL/kg	
Intradermal	0.05–0.1 mL/site	
Intramuscular	0.05 mL/kg/site	Maximum of 2–4 sites/animal.
Intranasal	50 µL	Manual restraint. Difficult to ensure that entire dose is in the nostril. Adverse effects unlikely but take care with fluids.
Intraperitoneal	10 mL/kg	
Intratracheal	40 µL	For a 200-g rat, use the method described for endotracheal intubation.
Intravenous	5 mL	Warm fluids to body temperature; administer over 2–3 min; single injection volume should not exceed 10% of the circulating volume; continuous 24 hr IV infusion should be ≤4 mL/kg/h.
Subcutaneous	5 mL/kg per site	Administer volume in 2 or 3 sites. Excludes Freund's adjuvant.
Subplantar	0.1 mL/foot	Subcutaneously on the plantar aspect of a hind foot; usually inject only 1 foot to not impair mobility.

material necessitating the use of 21–25-ga needles for administration (**Figure 69**). The use of a Hamilton syringe permits accurate delivery of very small volumes (1–50 µL). Once the needle is inserted, aspirate prior to injection to verify needle placement. Readers seeking more detailed information are referred to the review article by Morton et al. (2001) and the two articles by Turner et al. (2011).

Subcutaneous

This is usually performed over the dorsal and lateral sides of the rat (**Figure 70**). While tenting the skin, insert a needle between the skin and underlying tissues. If nothing is aspirated into the syringe, inject the compound. Redirect the needle or use other sites if injecting large volumes. Most individuals can incorporate this technique

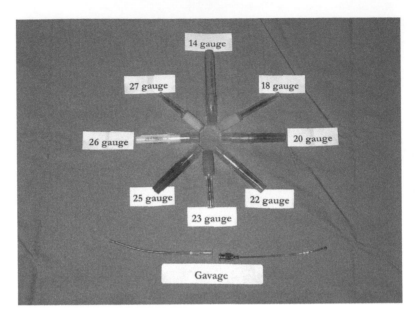

Fig. 69 Display of various needle sizes and types. Injectable needle cover colors are not standardized and may not be typical. Use of 22–25-ga needles is most common in rats.

Fig. 70 Subcutaneous injection of a compound to a rat.

into their restraint. Care must be taken by the individual administering the compound not to inject either him- or herself or the others restraining the rat. SC injection may be performed needle-free on adult rats using commercially available injection systems powered by a CO_2 cartridge (Biojector® 2000, Bioject Medical Technologies, Tualatin, OR).

Intraperitoneal

Administering compounds IP results in their absorption into the hepatic circulation prior to their distribution to other organs. Injection should be made just off the animal's midline in the lower right quadrant to avoid vital organs such as the cecum, ~72% of which is located on the left side of the abdomen. Restraining the animal in a head-downward position (the Trendelenburg position) to "push" the abdominal organs cranially also facilitates injection. The needle should enter the abdomen at a shallow angle (**Figure 71**). Aspiration of undesirable material (e.g., urine, bile, or gastrointestinal contents) indicates the needle is improperly placed and the compound contaminated; when undesirable material is encountered, euthanasia should be considered.

Intramuscular

The caudal hindlimb muscles are commonly used for rat IM injections. Alternatively, the quadriceps muscle group in the cranial aspect of the femur or the epaxial muscles are used, as they are devoid of major nerve and blood supply. The needle should be directed cranially if injecting into the quadriceps or caudally when injecting into the hamstrings, and inserted not too deep, as it is important to avoid hitting bone and the sciatic nerve, which runs along the caudal aspect of the femur. Injection into the nerve may cause discomfort and lameness; nerve irritation may result in rats chewing at their paws. Aspirate after needle insertion to ensure that a blood vessel is not penetrated.

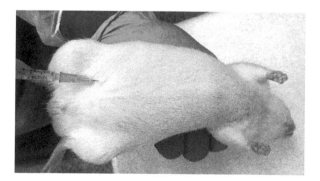

Fig. 71 Intraperitoneal injection of a compound to a rat. The rat's head is lower than the body and the injection is into the rat's lower right quadrant.

In a rat study, myotoxic drugs were lipid-soluble and most amphiphilic. The local injury after a single IM injection is fully reversible; however, frequent injections at the same site may cause permanent muscular damage.

Intravenous

IV administration sites include the lateral tail, dorsal metatarsal, jugular, femoral, sublingual, and penile veins. Typically a small quantity of blood will "flash" into the needle hub if properly positioned in the vein. Formation of a bleb or bubble indicates an extravascular injection. As in blood collection, one should start distally (farthest from the heart) and move proximally in case a repeat injection is necessary. Cleaning technique such as swabbing the injection site with alcohol is necessary to prevent infection. Sterile technique is required for catheter implantation for chronic compound administration. Upon needle or catheter removal one should apply adequate digital pressure to minimize hematoma formation.

IV injection techniques are similar to those of blood collection. The dorsal metatarsal vein is entered using a 25-ga needle. The jugular and femoral veins typically require cutdowns and a 25-ga needle for the injections, whereas the sublingual and penile veins require anesthesia and a 25- to 30-ga needle. Bolus injections require the compound to be compatible with blood and not too viscous. When large volumes are required, the injection material should be warmed to body temperature to minimize circulatory shock.

Intradermal

ID injections are typically performed for assessment of immune, inflammatory, or sensitization response. In rats, ID injections are given into shaved areas over the dorsal thorax and abdominal region. Position the needle at an acute angle to the skin while the skin is held taut so the bevel of the 23-ga or smaller needle enters only the skin and not the subcutaneous tissue. One may ensure proper placement upon injection when a small bleb forms.

Gastric Gavage/Per Os

Gastric gavage involves the use of a bulb-tipped gastric gavage needle. Prior to use, one should determine that the length of the needle is sufficient, i.e., approximately the distance from the mouth to

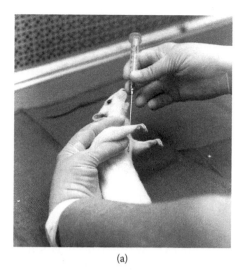

(a)

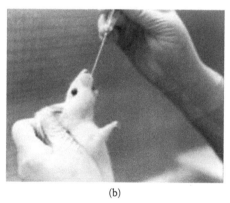

(b)

Fig. 72 Oral gavage. (a) Measure the bulb-tipped oral gavage needle before placement. (b) Gently but firmly advance the gavage needle starting from the side of the mouth; the rat will swallow the needle. Once positioned in the stomach, administer the compound.

the end of the sternum (**Figure 72**). Firmly restrain the animal in a vertical position and pass the needle through the side of the mouth, between the incisors and the premolars. Advance the needle towards the esophagus while rotating slightly and letting the animal swallow. Insertion in the trachea is likely if resistance is encountered. Administration of a small amount of the substance without vocalization or struggling indicates proper positioning.

Pneumonia may develop from an inadvertent administration into the lungs. Struggling during administration or excessive force in advancing the needle may also rupture the esophagus or pharynx. Should any of these complications occur, euthanasia should be considered. One may wish to practice the gavage technique with normal saline, which will be absorbed if a small quantity enters the lungs. Gavaging has previously been shown to stress rats acutely, with a significant correlation between heart rate and dosage observed until 10 minutes after gavaging.

Intracolonic

Colonic administration is a requirement for some pharmacokinetic studies. The material to be administered should be in a liquid or gel vehicle (e.g., 2% hydroxyethylcellulose), and administration volume should not exceed 3.5 mL/kg. While under gas anesthesia, wipe the rat's perianal region with saline, and slowly and gently insert a lubricated 16- to 18-ga gavage needle approximately 6 cm. While firmly holding the anal region, slowly inject the compound and then slowly withdraw the gavage needle. Observe for anal leaking. To prevent the animal from licking the perineal region and ingesting the compound, fit it with an E collar, which can be removed in 1–2 hours. Remove the animal from gas anesthesia and closely observe during recovery.

Intratracheal

Endotracheal intubation (Chapter 4), gastric gavage needles, and tracheostomy have been described for intratracheal administration. Whereas intubation and gastric gavage needles offer significantly less pain and distress, a tracheostomy provides a way to deliver solids/powders. When administering an intratracheal volume, it is important to administer a sufficient volume of air afterwards to displace the entire volume administered. Intubation has also been described for aerosol delivery.

Intranasal

This technique is used when studying compounds (including pathogens) absorbed through the respiratory mucosa. Dosing is nearly inaccurate, but the rat can be trained for the procedure. Using the intranasal route may result in elevated brain levels of compounds; this has been described for manganese and methotrexate.

Subplantar

One may administer up to 100 µL subcutaneously on the plantar aspect of the hind foot. In general, only one hind foot should be injected, since feeding requires the animals to use both hind feet to access the overhead feeders found in most cages.

Intraosseous

Intraosseous access for fluid administration and bone marrow transplant have been reported using a trochanteric fossa and a lateral femoral cutdown, respectively, under general anesthesia and using a surgical preparation of the site. When accessing via the trochanteric fossa, one may use a spinal needle or 20- to 22-ga needle. Alternatively, with the lateral femoral cutdown, a bur hole can be created with a 27-ga needle and the resulting femoral hole sealed with bone wax, followed by a two-layer closure.

Intracerebral, Intrathecal, Epidural

These techniques are usually performed in neuroscience, anesthesiology, and neurotoxicology studies, especially to deliver compounds to the central nervous system when the blood-brain barrier must be crossed, or when direct systemic effects are prevented. The techniques require rats to be sedated or anesthetized, aseptic skin preparation, the use of sterile technique for needle insertion, and well-trained personnel.

Intracerebral injections are most commonly performed using stereotactic surgery (see below) to ensure precise, consistent delivery. Death may occur from administering substances with a high protein content or excessive volume, or too rapidly. Dyes (e.g., Evan's blue) are reported to localize the injection site postmortem.

Intrathecal (subarachnoid) injections require puncturing the dura, and thus subarachnoid space catheterization. The original technique involved inserting a catheter through the atlanto-occipital membrane and advancing it to the lumbar subarachnoid space (Yaksh and Rudy 1976; LoPachin et al. 1981). Alternative techniques include approaches to the L4–L6 vertebrae, (Xu et al. 2009; Størkson et al. 1996). Continuous infusion via a T13–L2 laminectomy and placing an osmotic mini-pump in a subcutaneous pouch is described by Guazzo et al. (1995). Visualization of cerebrospinal fluid (CSF) after needle insertion confirms correct placement of the needle.

Direct transcutaneous intrathecal injections can be performed at the L4–L6 intervertebral space using a 30-ga needle attached to a small-volume syringe (e.g., 10-μL Hamilton syringe), as described by Schwartz et al. (2009) and Yowtak et al. (2011). The rat can be placed in a device similar to that shown in **Figure 73** to facilitate the injection.

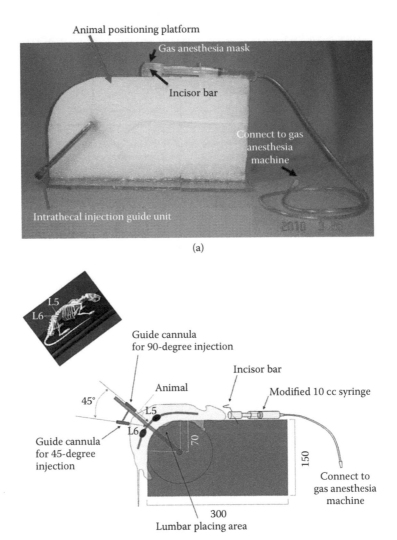

(a)

(b)

Fig. 73 Intrathecal injection guide (a) and schematic (b). (Image courtesy of Dr. J.M. Chang, University of Texas Medical Branch.)

Epidural injections differ from intrathecal injections, especially regarding the kinetics of substance absorption, which can be markedly different. With this technique, the compound is placed in contact with, but not entering, the dura. If CSF is noted when attempting the technique, the needle should be withdrawn and repositioned. Alternatively, reduce the dose of the administered compound.

Osmotic Pumps

Osmotic pumps deliver compounds continuously and reliably at controlled rates over a given interval. These implantable pumps can deliver compounds IP, SQ, IV, intracerebrally, or intra-arterially. One may use a pump to target delivery to a specific tissue or organ. Pumps are available in a wide variety of delivery rates and intervals. The pumps function through an osmotic gradient between the pump and the surrounding tissue.

other techniques

Physiological Recording

Physiological recording is a powerful data collection tool, especially when coupled with telemetry to permit data collection and transmission in an environment that either consists of or approximates the animal's home cage. Telemetry systems used in rats are primarily implantable (e.g., temperature, activity, electroencephalogram, electrocardiogram, blood pressure), while external jacketed systems are available for larger animals. Depending on the system employed, battery life must be a consideration and frequently consists of weeks to months.

Electroencephalogram

EEG provides a continuous tracing of regional brain electrical activity and is usually employed in sleep research and as a supplement to safety pharmacology evaluation. EEG may also assist in behavior or brain electrophysiology studies. Several techniques have been described that permit either free movement (telemetry) (Tang et al. 2007; Fitzgerald et al. 2003; Petty 1982) or tethering (cabled system) (Gohd et al. 1974; Petty 1982). An example of a protocol for implanting EEG electrodes including stereotaxic coordinate details is available online (NeuroDetective International: www.ndineuroscience.com/ pdf_pubs/Rat_Sleep_EEG_Methods.pdf).

Electrocardiogram

ECG is important in cardiac toxicology and transplants and in other studies where information on the heart's electrical activity is vital. The use of three-dimensional ECG and cardiac-gated magnetic resonance microscopy are other valuable research tools. Telemetry is the gold standard for ECG recording and acquisition of hemodynamic signals in pharmacology/pharmacokinetics studies; however, the implementation of telemetric systems entails high costs. Alternative methods include implanting electrodes subcutaneously as described by Robineau (1988) and fitting a custom-made elastic cotton jacket with two pieces of electrodes attached on its inner surface to the rat's mean thoracic circumference. Reader are directed to Martín Barassa et al. (2008), Petty (1982), and Sambhi and White (1960) for techniques on anesthetized animals.

Blood Pressure

Radiotelemetry is the preferred method for automatic collection of chronic, continuous BP data. Rat telemetry implantation involves advancing a BP catheter into the abdominal aorta and placing a radio-transmitting device in the peritoneal cavity. Besides the aorta, one may place a catheter in the carotid or femoral artery, exteriorize the catheter, and connect it to a pressure transducer. Refer to Huetteman and Bogie (2009) for details. Tail cuff blood pressure measurements can be performed in conscious rats by using commercially available noninvasive systems.

Diagnostic Imaging

Employing diagnostic imaging modalities permits researchers to use fewer animals and extract additional information in a study, and may offset the expense of these imaging techniques. One may follow a single animal over a period and study dynamic biological processes in intact small-animal models instead of sacrificing many animals over the same period. Maximize results and minimize problems associated with bioimaging (e.g., stress and hypothermia) by careful procedural planning and execution. These include animal preparation ranging from simple anesthesia (for computed tomography imaging of bone morphology) to the more complicated involving fasting, anesthesia, and/or probe injection (e.g., for metabolic imaging with positron emission tomography tracers). Furthermore, animals may be radioactive or exposed to biological vectors (e.g., viral vectors) in the imaging process, thereby necessitating a close working relationship with knowledgeable and experienced biosafety professionals. Interested readers are referred to the *ILAR Journal* issue 2008: 49(1), *Noninvasive Bioimaging of Laboratory Animals.*

Ultrasound

Cardiac imaging is one of the most common uses of ultrasound in laboratory animal research. Echocardiography is used to evaluate ventricular function in mice and rats, using techniques and indices from human echocardiography. M-mode images enable the measurement of wall thickness, left ventricular (LV) dimensions, and cardiac mass, while combined B- and M-mode images show contractility, LV dilation, and fractional shortening.

Ultrasound may be used for rat pregnancy detection. Transabdominal ultrasonography using a 12-MHz linear transducer can detect embryonic vesicles at day 9 of pregnancy. By day 12, crown-rump length can be measured, and heart rate can determined by pulsed-waved color Doppler sonography.

Ultrasound imaging is further described in Chapter 4. It is necessary to have good personnel to interpret the image and ensure reliable measurements from the images, which have lower image contrast than CT or MRI images.

Radiography

Flat film/digital

Flat-film conventional radiography is briefly discussed in Chapter 4. Digital radiography is available and can provide high-magnification images by using digital x-ray sensors instead of traditional photographic film. Other digital radiography advantages include time efficiency (bypasses chemical processing) and the ability to digitally transfer and enhance images. Systems that include gas anesthesia are commercially available for imaging mice, rats, tissue samples, and successfully detect bone lesions related to ischemia and thrombosis. Such systems, nor fluoroscopy, were as sensitive as bioluminescence imaging in detecting vertebral metastases in a rat osteolytic model.

Fluoroscopy

Fluoroscopy is more commonly used for surgical procedures and produces real-time imaging. It is frequently used in conjunction with contrast agents, such as barium to find gastrointestinal abnormalities and in gastrointestinal physiology studies (e.g., transit time) and iodine used in angiography and visualization of extrahepatic biliary anatomic structures and function.

Computed tomography (CT)

In CT, an x-ray source and detector rotate around the specimen, producing a three-dimensional serial image and volumetric data with anatomical information; differential tissue x-ray absorption generates contrast. Small animal CT scans can achieve a very high resolution, in the range of a few micrometers. However, CT is less sensitive than PET, requiring contrast agent administration (e.g., iodinated compounds) at higher doses; contrast agents may cause pharmacologic effects in the animals.

Dual-energy x-ray absorptiometry (DEXA)

DEXA uses two x-ray beams of differing energy levels to measure body composition and bone mineral density (BMD). In small animals, DEXA has been performed mainly on bone specimens. In one study, bone mineral ratio, as determined by BMD, measured by DEXA proved a practical index in evaluating laboratory animal bone mineral characteristics.

Bioimpedance spectroscopy (BIS)

Considered the method of choice for body composition, BIS provides rapid results with portable equipment. Ward and Battersby (2009) compared BIS and DEXA in assessing rat body composition.

Optical imaging

Optical imaging uses either fluorescence or bioluminescence to generate and detect visible light in the body. The fluorescent method uses fluorophors, which can emit photons after excitation. Fluorescent reporter proteins such as green fluorescent protein (GFP) and red fluorescent protein (RFP) are used widely for cell labeling and gene expression monitoring. Bioluminescent methods use an enzymatic reaction between the luciferase enzyme and its substrate, a luciferin, to produce light for detection. The most common bioluminescent reporter proteins for imaging in living small animals are the firefly luciferase from *Photinus pyralis* and the *Renilla* luciferase from the sea pansy *Renilla reniformis*. One key technical difference between fluorescence and bioluminescence imaging is that the latter requires substrate injection into the animal prior to imaging, whereas no additional animal manipulation is necessary for fluorescence imaging of GFP or RFP expression. Most optical imaging systems can use either bioluminescence and fluorescence, and some even have an optional x-ray capability.

Magnetic resonance imaging (MRI)

MRI evaluates anatomical, physiological, and chemical processes. It uses strong magnetic fields to align the spin of hydrogen nuclei in tissues. The application of a radio waves pulse "flips" the spin of the hydrogen nuclei into an excited state. The subsequent relaxation of the hydrogen spin, realigning with the principal magnetic field, emits a pulse of radio waves that are detected and quantified. MRI provides a much higher contrast than CT between various soft tissues, and paramagnetic contrast agents can be employed to accentuate this contrast.

Positron emission tomography/single photon emission computed tomography (PET/SPECT)

Like MRI, PET and SPECT evaluate anatomical, physiological, and chemical processes. PET and SPECT rely on photon detection from radiolabeled probes. PET assembles the coincident detection of high-energy gamma rays emitted from positron-labeled molecules in the subject, creating tomographic images. In contrast, SPECT supplies tomographic information by rotating detectors around the subject and collecting only gamma particles hitting detectors through a collimator (an aperture permitting particle detection only of those particles traveling directly through the aperture). A major advantage of PET/SPECT is its high sensitivity and the ability to quantify the concentration of imaging probes in the body. A typical small-animal PET scan typically ranges from 5 minutes to 1 hour, entailing the IV or IP injection of a positron-emitting imaging probe.

Hybrid imaging

Hybrid imaging provides combined imaging modalities in a single device. Examples include PET with CT, and PET with MRI. The resulting device is a powerful combination of functional/

molecular (PET) and anatomical (CT or MRI) data into a single scan.

Synchrotron

A synchrotron accelerates charged particles along a track by using electric and magnetic fields. The track may be constructed of straight and curved sections. While the electric field propels the particles, the magnetic field "guides" the particle beams around the track. Synchrotrons are not all alike, with the magnetic field strength and track radius determining the maximum energy imparted. Conventional full-sized (e.g., football field-sized) synchrotrons require a significant capital investment; as a result, there are only about 50 such facilities worldwide. Compact synchrotron light sources are available and provide x-ray sources on a significantly smaller scale than full-sized, conventional synchrotrons.

As charged particles in the synchrotron lose energy, they generate photons over the electromagnetic spectrum, including the x-ray, ultraviolet, and infrared regions. Select electromagnetic spectral frequencies are separated and pass through beamlines to a procedure area where various analyses occur (e.g., crystallography, tumor and tissue imaging, biologic processes/drug interactions).

Gait Analysis

Gait analysis is beneficial for a wide variety of arthritis, toxicology, peripheral nerve damage, and neurodegenerative models. Gait analysis is modeled after studies performed by Parker and Clark (1986, 1990) that evaluated gait topography and normal locomotion. Analyses include stride width, stride length, and velocity. Stride width and length may be measured by applying stamp-pad ink to the hindpaws and then having the animals walk through an opaque tunnel (1 m long × 7 cm high × 10 cm wide) with a tunnel floor lined with white paper. Automated gait analysis systems are also available to measure and record various rat gait characteristics using treadmill locomotion.

---

common research uses of rats

The laboratory rat has many common research uses. Whatever the study, it is important to minimize variables that may confound the research data. For example, genetic variability occurs in stocks and strains from different suppliers. Therefore, researchers must be familiar with the colony history and breeding dynamics to assess the degree to which a sample genotype will be representative of the whole species. Genetic monitoring for individual loci includes biochemical or immunological markers, DNA restriction length polymorphisms, and coat color markers. Skin grafting, polyvalent alloantisera, DNA fingerprinting, morphology, and breeding performance are methods to monitor several loci concurrently. Infection is another variable that could have catastrophic effects on research; murine adventitious pathogens are not a component of the animal modeling process. Other variables include sex, nutrition, and husbandry. All these factors must be considered in light of the experimental paradigms and goals.

Aging Studies

The laboratory rat is useful for aging studies, whether the questions are related to toxicology, neoplasia, or aging, because of its short lifespan, moderate size, and low maintenance cost.

- Rodent diets should be free of contaminants and nutritionally complete. Caloric restriction after 6 months of age increases the lifespan of rats. Diet can also influence the onset of age-related pathology such as chronic glomcrulonephropathy.
- Aging rodent colony husbandry should take into consideration infectious disease prevention through a barrier system and reducing environmental stresses such as noise and animal density.
- The Fischer 344 rat, the Brown Norway rat, and their F1 hybrid are the three most commonly used inbred rat strains for aging studies in the United States. The Sprague-Dawley, Wistar, and Long-Evans outbred stocks are also common.

Oncology, Toxicology, Teratology

These three areas share similar methods and principles. Along with mice, rats are the most common species used in studies identifying the carcinogenic, toxic, and teratologic potential of chemicals and drugs. In addition, they are frequently used to study the carcinogenic process, both tumor initiation and promotion. Rat models of various cancers also are important in cancer therapy development.

- Experimentally induced rat tumors are those arising in various organs following administration of known carcinogens such as 7,12-dimethylbenz(a)anthracene (DMBA).

- Spontaneous tumors are those arising during the natural course of a rat's lifetime. The incidence of spontaneous tumors varies with rat age, strain, and sex.

- Model selection depends on the nature of the oncology study. A model producing a tumor in a target tissue with high relevance to human cancer may be most useful in a chemotherapy trial. On the other hand, models with spontaneous tumors that are less common in humans may be the better choice for understanding the basic mechanisms of tumor initiation or promotion.

- Rats have long been used as a model for toxicity testing of various agents and for evaluating new drug safety and efficacy.

- The National Toxicology Program (NTP) establishes detailed specifications for experimental bioassay protocols evaluating the toxic and carcinogenic potential of chemical, biological, and physical agents.

- The rat is useful in teratologic studies because of its short reproductive cycle, large litter size, and relatively few spontaneous congenital anomalies.

- The route of administration and dose of a potential teratogen can influence its effect on the fetus. Potential teratogens are administered to the dam at specific time points during gestation. The fetuses are removed about 24 hours before birth to prevent cannibalism by the dam. Embryos are methodically examined grossly and histologically for evidence of malformations.

- Common rat stocks and strains used in toxicological, carcinogenic, and teratologic studies are Sprague-Dawley, F344, Wistar, and Long-Evans.

Cardiovascular Research

The rat is commonly used in cardiovascular (CV) studies despite cardiac anatomical, physiological, and pharmacological differences to humans and other higher mammals. The differences include rats having a left superior vena cava in addition to a right, rats' myocardial cells having a membrane action potential without a plateau, and rats being very resistant to digitalis glycosides. Therefore, understanding the advantages and disadvantages of a rat CV model is necessary in extrapolating results to humans.

Transgenic hypertensive and induced CV rat models are available. Readers are referred to the reviews by Bader (2010) and Doggrell and Brown (1998) for further information. Popular rat CV models include the following:

- Spontaneously hypertensive rats (SHR), and substrains, are well-characterized models, with the Wistar Kyoto rat (WKY) serving as a normotensive control. SHR are pre-hypertensive for the first 6–8 weeks of age with systolic blood pressures (BP) of 100–120 mmHg. Hypertension (BP of >150 mmHg) develops over the next 12–14 weeks of age. Males are more severely affected than females.

- The development of hypertension and heart failure in the Dahl/Rapp salt-sensitive (SS) rat can be controlled by titrating dietary salt. The Dahl/Rapp salt-resistant (SR) rat is the normotensive control strain.

- Zucker rats develop borderline hypertension and cardiac hypertrophy secondary to noninsulin-dependent diabetes and obesity.

- Rats are widely used in myocardial structure and metabolism studies, both *in vivo* and *in vitro*.

- Telemetry devices have greatly facilitated and remain the gold standard in measuring cardiovascular parameters.

Behavior and Neuroscience

Rats are probably the most widely used mammal in neuroscience research. They are easily trainable and motivated to perform designated tasks with either positive or negative reinforcement. The National Institutes of Health (NIH) handbook *Methods and Welfare Considerations in Behavioral Research in Animals* (2002) provides a description of and references for commonly used behavioral research methods and associated animal welfare considerations. The *Guidelines for the Care and Use of Mammals in Neuroscience and Behavioral Research* (2001) (the "Red Book") also provides information on best use practices for mammalian species in neuroscience research.

- Spontaneous rat models include the alcohol-preferring AA (Alko alcohol) and alcohol-avoiding ANA (Alko nonalcohol) rats for alcohol abuse studies; WAG/Rij and the genetically epilepsy-prone rats (GEPR/3 and GEPR/9) for epilepsy studies; and SHR rats for attention-deficit hyperactivity disorder (ADHD) studies.

- Induced models include the intranasal or intrastriatal administration of 1-methyl-4-phenyl-1,2,3,6-tetrahydropyridine (MPTP) for Parkinson's disease research; and limiting access to fatty diet for studies of intermittent excessive behaviors such as binge eating.

- Several testing paradigms have been used in rats. These include the elevated T-maze to study anxiety, Morris water maze to study learning and memory, and the open field apparatus to study exploratory behavior. When such equipment are shared by different laboratory groups, equipment sanitation and disinfection is important to prevent cross-contamination when a disease outbreak occurs. Crawley (2003) outlines a rat behavioral phenotyping strategy.

- There are strain, sex, and age differences in behavior, learning and memory, and motor function. Strain differences are reported for various water rewards in operant conditioning testing.

- Environmental enrichment can affect neuroscience research outcomes. Interested readers are referred to the review by Simpson and Kelly (2011). IACUCs require scientific justification for not providing animals environmental enrichment.

Stereotactic surgery

Stereotactic surgery is commonly used in neuroscience research, especially to implant microelectrodes or microprobes in specific structures of the brain. The positions of the brain nuclei are determined as the distance from a defined reference point on the skull, principally the bregma, the point at which the coronal suture is intersected perpendicularly by the sagittal suture. Other important points include the lambda (intersection between the sagittal and lambdoid sutures) and the interaural midpoint (an imaginary line that passes through the brain at the midpoint of the external auditory meatus). Paxinos et al. (1985) indicated that stereotaxic atlases could be used with rats of different sex and strain provided that the rats' body weights conform to those used in the reference atlas. In rats of different weights, accuracy is greater if bregma and the interaural line are used as reference points for work with rostral and caudal structures, respectively.

It is crucial to correctly locate the bregma. This position could be roughly determined, but this often results in error. Such errors can be due to the landmark not being clear for some animals, and the differences between individuals in the shapes of sutures and the alignment between the skull and the brain. In this regard, other systems have been developed to locate the bregma. Examples include analyzing digital pictures of the exposed skullcap with a computer, and using a new coordinate system with the origin being the intersection point of the posterior edges of the two cerebral hemispheres. Paxinos and Watson (2007) discuss in detail stereotaxy in rats.

Biodefense

Laboratory rats are used to study various aspects (e.g., pathogenesis, treatment) of biodefense models including plague, Rift Valley fever, and ricin poisoning. Interested readers may refer to the book on this subject edited by Swearengen (2005).

other rats used in research

Listed below and summarized in **Tables 30** through **40** are less commonly used research rats. They have inherent characteristics that make them valuable research animals. These rodents are typically unavailable commercially, with institutional breeding colonies frequently serving as sources. It is important to note that these rats, together with wild-caught *Rattus* spp., are covered by the United States Department of Agriculture (USDA). The *Guidelines of the American Society of Mammalogists for the Use of Wild Mammals in Research* provides current professional techniques and guidance involving mammals used in research and teaching, and includes details on marking, housing, trapping, and collecting animals.

Kangaroo Rat (*Dipodomys* spp.)

Features include:

- Approximately 20 species found in North America.
- Adapted to bipedal locomotion, with the tail used for balance.

TABLE 30: REPRODUCTIVE VALUES FOR *DIPODOMYS* SPP.

Parameter	Value/Range
Sexual maturity	2 mo
Breeding season	Year-round
Litters per year	3
Gestation	29–33 d
Litter size	3.5 pups
Weaning	21–24 d
Estrous cycle length	6 d

- Seldom drink, but produce water from the breakdown of food.
- Kidneys have four times the concentrating capacity of human kidneys.
- All teeth grow throughout life.
- Nonreceptive females are very aggressive and reportedly will kill males.
- Kangaroo rats possess fragile tails that will break if used for animal restraint; for proper restraint one should firmly grasp the skin on the dorsum of the neck and restrain the hind limbs.

Husbandry:

- Primarily seed eaters
- Require dust baths
- No water ad libitum (water addiction)

Research uses:

- Renal physiology and water conservation
- Whole body irradiation
- Psychotropic drug effects

Indian Soft-Furred Rat (*Millardia meltada*)

Features include:

- Animals are slow-moving and easy to handle.
- Young are born with thin hair and well-grown incisors.

TABLE 31: NORMAL VALUES FOR *MILLARDIA MELTADA*

Parameter	Value/Range
Adult weight	84.5–100 g
Lifespan (usual)	6.2 mo
Lifespan (maximum)	16 mo
Chromosome number	2n = 50
Vagina opens	35–66 d
Gestation	20 d
Average litter size	4.09 pups
Eyes open	12 d
Mean blood pressure	95.1 ± 14.5 mmHg

- Young attach to the mother's teats with their incisors.
- X chromosome is large and metacentric.

Research uses:

- Parasitic infections
- Androgen-dependent mammary tumors
- Reproduction studies

Multimammate Rat (*Praomys coucha*)

Features include:

- Widely distributed throughout Africa.
- Y, Z, and GRA-Giessen strains have been described.
- No gall bladder.
- Submaxillary glands are the richest known source of nerve growth factor.
- Females have 8–18 pairs of mammary glands, a well-developed prostate, and a strong cannibalistic tendency with the first litter; clitoral gland is absent.
- Males lack preputial glands.

Husbandry:

- Monogamous pairs should be established.
- Seed eater in the wild but thrives on commercial rodent feed.
- Animals will bite without provocation.

TABLE 32: NORMAL VALUES FOR *PRAOMYS COUCHA*

Parameter	Value/Range
Adult weight	40–80 g
Lifespan (usual)	2–3 yr
Lifespan (maximum)	38 mo
Food consumption	6 g/d
Chromosome number	2n = 36
Body temperature	♂: 35.9°C; ♀*: 36.9–37.5°C
Puberty, ♂ and ♀	55–75 d
Breeding season	Year-round
Estrous cycle	6–8 d
Postpartum estrus	Yes
Litter per year	Several
Gestation	23 d
Litter size	5–20 pups
Birth weight	2–3 g
Eyes open	13–17 d
Weaning	19–21 d

* Female temperatures are lactation dependent: nonlactating females, 36.9°C; lactating females, 37.5°C.

TABLE 33: HEMATOLOGICAL VALUES FOR *PRAOMYS COUCHA*

Parameter	Value/Range
WBC count (total $\times 10^3/\mu L$)	2.8–13.0
Neutrophils (%)	8–48
Lymphocytes (%)	48–93
Monocytes (%)	0–1
Eosinophils (%)	0–9
Basophils (%)	0–1
Platelets ($\times 10^3/\mu L$)	208–754
PCV (%)	40
RBC ($\times 10^6/mm^3$)	7.5
Hemoglobin (g/dL)	13.0

Research uses:

- High incidence of neoplastic and pre-neoplastic lesions
- Thyroid disease
- Gastric/duodenal ulcers
- Gastric carcinoid (Zollinger-Ellison syndrome)
- Degenerative joint disease
- Lassa virus studies—the only nonhuman natural host

White-Tailed Rat, South African Hamster (*Mystromys albicaudatus*)

Features include:

- Found in Central and South Africa.
- No cheek pouches.
- Large ventral sebaceous gland.
- No gall bladder.
- Females have a rudimentary prostate and two pairs of inguinal mammae.
- Males have an os penis.
- Newborns attach to the mammary glands for 2–3 weeks.

Husbandry:

- Establish breeding pairs at a young age and maintain as monogamous lifetime mates.
- Remove the male prior to parturition.
- The tail is too fragile for lifting, so lift by the thorax.
- High mortality associated with the application of polymyxin B sulfate, bacitracin, and neomycin sulfate triple antibiotic ointment) for 2 weeks.

TABLE 34: NORMAL VALUES FOR *MYSTROMYS ALBICAUDATUS*

Parameter	Value/Range
Adult weight: ♂/♀	85/130 g
Lifespan (usual)	2.4 yr
Lifespan (maximum)	6.2 yr
Water consumption	5 mL/d
Chromosome number	2n = 32
Puberty: ♂/♀	4–7/4–5 mo
Breeding season	Year-round
Estrous cycle: range	4–9 d
Postpartum estrus	Yes
Litters per year	Several
Gestation	36–39 d
Litter size	1–5 pups
Birth weight	5.0–7.8 g
Eyes open	16–20 d
Weaning	25 d

Research uses:

- Diabetes mellitus
- Dental caries/periodontal disease
- Infectious disease (especially American cutaneous leishmaniasis caused by *Leishmania braziliensis*)
- Radiation research

Degu, Trumpet-Tailed Rat (*Octodon degu*) (Figure 74)

Features include:

- Found in South America.
- Cheek teeth have a peculiar deep fold resembling the number 8.
- Cervical and thoracic thymus.
- Left and right anterior vena cava.
- Dental formula: 2(1/1 0/0 1/1 3/3) = 20.
- Diurnal and active throughout the year.
- Young are born fully furred with eyes open (precocious).
- Enhanced metabolism of morphine and pentobarbital.
- A stereotaxic atlas of the brain is available (Wright and Kern 1992).
- Females are induced ovulators, and have 8 mammae and a vaginal closure membrane.
- Males have intra-abdominal testicles.

Fig. 74 An adult female degu (right) with young (left). (Image courtesy Dr. Sharon Vanderlip.)

TABLE 35: NORMAL VALUES FOR *OCTODON DEGU*

Parameter	Value/Range
Adult weight	200–300 g
Litter per year	2
Gestation	90 d
Litter size	5–6 pups
WBC count (total $\times 10^3/\mu L$)	8.5 ± 0.39
Neutrophil/lymphocyte ratio	40:60
PCV(%)	42.1 ± 0.59
RBC ($\times 10^6/mm^3$)	8.69 ± 0.19
Hemoglobin (g/dL)	12.0 ± 0.15
Total protein (g/dL)	5.7 ± 0.2

Husbandry:

- Can climb trees, possible source of environmental enrichment.
- Requires dust baths twice per week.
- Do not handle by the tail; scoop up with the tail as the skin may slip from the tail.
- Feed commercial rodent chow supplemented with vegetables (not *ad libitum*, to prevent obesity).

Research uses:

- Diabetes mellitus
- Reproduction and neurobiology
- Thymus research
- Ocular pathology

Fat Sand Rat (*Psammomys obesus*)

Features include:

- Found in Northern Africa and the Middle East.
- Prefers leaves and stems from plants of the family *Chenopodiaceae* as food items. These plants have a high concentration of salt and water.
- Kidneys produce highly concentrated urine.
- Communicates through foot thumps and high-pitched squeaks.

TABLE 36: NORMAL VALUES FOR *PSAMMOMYS OBESUS*

Parameter	Value/Range
Adult weight	109–276 g
Lifespan (usual)	3 yr
Breeding season	Year-round
Gestation	23–25 d
Litter size	2–5 pups

- Nongrooved incisors.
- Secretes more insulin than most other species
- Commercially available (Harlan Laboratories).

Husbandry:

- Feed: 5 g of commercial rodent ration with 50 g of beets and spinach per animal per day. Also provide vitamin D or a general vitamin/mineral supplement.
- Water: should consist of a 1.5% (15 mg/mL) sodium chloride solution.
- Best reproductive ability with 14 hours of light.

Research uses:

- Diabetes mellitus type II
- Renal physiology
- Reproduction

Cotton Rat (*Sigmodon hispidus*) (Figure 75)

Features include:

- Found in the southern United States, Central and South America.
- S-shaped molar cusps.
- Young are precocious.
- Permanent monogamous pairing by 3–4 weeks of age (at weaning) is most successful; pairing of older animals increases the likelihood of fighting.
- Good foster mothers (readily accept foster pups, even if they are a different age than their own litter).
- Not aggressive, but will attempt to bite when picked up.

Fig. 75 The cotton rat. (Image courtesy Harlan Laboratories.)

TABLE 37: NORMAL BIOLOGY VALUES FOR *SIGMODON HISPIDUS*

Parameter	Value/Range
Adult weight: ♂/♀	70–200 g
Body temperature	36.5±0. 1°C (97.7±0.2°F)
Chromosome number	2n=52
Lifespan (usual)	23 mo
Lifespan (maximum)	3 yr
Puberty: ♂/♀	30–50 d
Breeding season	Year-round
Estrous cycle (usual)	9 d
Estrous cycle (range)	4–20 d
Postpartum estrus	Yes
Litters per year	Several
Gestation	26–28 d
Litter size	2–10 pups
Birth weight	7.09 g
Eyes open	1 d
Weaning	21 d

- Moves very fast and can jump vertically over 30 cm. Use a tube for transferring between cages.
- Commercially available (Harlan Laboratories).

Research uses:

- Dental caries
- Only animal model for human epidemic keratoconjunctivitis secondary to ocular adenovirus infection
- Respiratory syncytial virus, parainfluenza type 3, and human adenovirus 5 studies

TABLE 38: HEMATOLOGICAL VALUES FOR *SIGMODON HISPIDUS*

Parameter	Value/Range
WBC count (total × 10³/μL)	4.3–11.9
Neutrophils (total × 10³/μL)	2.8–9.2
Lymphocytes (total × 10³/μL)	3.2–5.9
Monocytes (total × 10³/μL)	0.1
Eosinophils (total × 10³/μL)	0.3–0.5
Basophils (total × 10³/μL)	0.1
Platelets (total × 10⁶/μL)	5.4–8.0
PCV (%)	33.3–43.5
RBC (× 10⁶/mm³)	3.9–6.2
Hemoglobin (g/dL)	10.5–17.1
MCV(mm³)	62.0–71.3
MCH (pg)	19.5–27.8
MCHC (g/dL)	31.3–32.6

TABLE 39: CLINICAL CHEMISTRY VALUES FOR *SIGMODON HISPIDUS*

Parameter	Value, Female*	Value, Male*
Total protein (g/dL)	7.06±0.34	7.09±0.62
Albumen (g/dL)	2.95±0.24	3.08±0.29
Blood urea nitrogen (mg/dL)	22.06±3.34	21.21±2.53
Creatinine (mg/dL)	0.65±0.18	0.50±0.12
Uric acid (mg/dL)	1.09±0.75	0.73±0.28
Total bilirubin (mg/dL)	0.18±0.13	0.19±0.19
Glutamate oxaloacetate transaminase (IU/L)	153.39±28.71	120.41±56.20
Glutamate pyruvate transaminase (IU/L)	82.88±33.96	63.95±34.39
Alkaline phosphatase (U/L)	123.45±31.20	139.19±29.82
Creatine phosphokinase (IU/L)	236.16±135.09	167.15±104.99
Lactate dehydrogenase (IU/L)	260.00±73.87	231.46±60.19
Glucose (mg/dL)	191.53±65.97	164.46±35.44
Cholesterol (mg/dL)	66.73±22.95	80.21±20.04
HDL cholesterol (mg/dL)	34.98±16.56	42.93±13.42
Triglyceride (mg/dL)	156.52±89.17	91.05±30.61

* mean ± standard deviation.

Naked Mole-Rat (*Heterocephalus glaber*)

Features include:

- Native to East Africa.
- Few pale-colored hairs scattered throughout the body and tail; prominent vibrissae; fine hairs along the edges of the feet; wrinkled skin is pinkish or yellowish.

TABLE 40: NORMAL VALUES FOR *HETEROCEPHALUS GLABER*

Parameter	Value/Range
Adult weight (size)	29–39 g (4–5 in. long)*
Chromosome number	2n=60
Body temperature	29–35°C (84–95°F)
Breeding season	All year
Estrous cycle	~30 d
Postpartum estrus	Yes (14 d after parturition)
Gestation	66–74 d
Litter size	1–29 pups (average=12)
Birth weight	1.86±0.33 g
Eyes open	~30 d
Wean	Starts at 3–4 wk of age

* Reached at ~2 yr of age.

- Lacks sweat glands and subcutaneous fat.

- Burrows 0.5–2.5 m below the soil surface.

- Lowest mass-specific metabolic rate for mammals as a physiological adaptation to underground living: low rates of gas exchange, heat production, and body temperature; tolerance of hypoxia and hypercapnia, vitamin D deficiency with vitamin D–independent calcium metabolism, streamlined body shape, funky walk, and small eyes.

- Longest-living rodent. Maximum lifespan exceeds 28 years in captivity and 17 years in the wild.

- Negligible senescence: Characterized by very slow or nonexistent changes in physiological parameters with age, including no age-related increase in mortality rate, changes in basal metabolism, body composition, or bone mineral density.

- Reproduction: Eusocial structure with a single breeding queen usually found atop a pile of huddling colony members; females become sexually mature only when breeding opportunities arise (e.g., when a breeding female dies or is removed from the colony); queen mates for life with the male to whom she was initially paired, but she also takes 1 to 3 of her sons as additional sires; prolific until death; offspring receive extended care, not only from the queen who nurses them but also from siblings who collect and carry food to the nest.

- Sexing: External genitalia of both sexes, except the breeding female, are very similar. Identify females by the presence of thin, often darker, red line between the anus and clitoris. Breeding female has a perforate vagina and prominent mammae.

- Generally healthy; diseases include dry, scaly skin (low humidity), *Escherichia coli* infections (abdominal distention and respiratory distress), oral thrush, fight wounds, broken incisors and abnormal incisor growth.

- Once an animal is out of the colony (as an escapee, or sick and isolated), it will be difficult to return it. Damping the animal with materials from the "toilet chamber" may be effective in avoiding aggression from colony members.

- A stereotaxic brain atlas is available (Xiao et al. 2006).

Note: Exposing naked mole-rats to human herpes simplex virus type 1 (HSV-1) may be fatal.

Husbandry:

- Consider the animal's physiology, behavior, and activity in the wild for housing needs.
- Maintained at 28–32°C and 50–65% relative humidity; constant darkness with supplemental heat/light by 25-W red/yellow light bulbs.
- Incorporate nesting, urinating, and defecating chambers.
- Feed vegetables (e.g, sweet potato, corn) to meet the daily water requirements.
- Animals begin eating solids at ~21 days but may continue nursing until 6 weeks of age.
- Avoid feeding rodent chow as the high and excess calcium can cause metastatic calcification.

Research uses:

- Physiological mammalian systems with variable body temperature (such as may occur during clinical hypothermia)
- Cancer (extraordinarily resistance to cancer)
- Pain research (skin lacking in *substance P*, a neurotransmitter responsible for sending pain signals to the central nervous system in mammals)

resources

organizations

Professional organizations serve as initial contacts for obtaining information regarding specific issues pertaining to the care and use of laboratory rats. One should consider membership at the personal and institutional level, as it permits the laboratory animal professional to stay abreast of regulatory issues, improved animal procedures and techniques, management and personnel issues, and animal health issues. Although contact details were unavailable for some organizations, the organizations' names have been included in this listing.

Academy of Surgical Research (ASR)
Website: www.surgicalresearch.org

African Biological Safety Association (AfBSA)
Website: http://afbsa.org
Facebook: www.facebook.com/group.php?gid=56691104667

Agri-Food and Veterinary Authority (AVA, Singapore)
Website: www.ava.gov.sg

Alternatives to Animal Testing at the Johns Hopkins Bloomberg School of Public Health (AltWeb)
Website: altweb.jhsph.edu
Facebook: www.facebook.com/JHUCAAT?ref=ts
Twitter: twitter.com/jhucaat

American Association for Laboratory Animal Science (AALAS)
9190 Crestwyn Hills Dr., Memphis, TN, USA 38125-8538
Telephone: 901-754-8620
Fax: 901-753-0046
Email: info@aalas.org
Website: aalas.org
Facebook: www.facebook.com/pages/AALAS/221279445386
Twitter: twitter.com/AALASnational

American Association for the Advancement of Science (AAAS)
Website: www.aaas.org
Facebook: www.facebook.com/AAAS.Science
Twitter: twitter.com/AAAS_News

American Association of Pharmaceutical Scientists (AAPS)
Website: www.aapspharmaceutica.com
Facebook: www.facebook.com/group.php?gid=12049321194
Twitter: twitter.com/AAPSComms

Animal Behavior Management Alliance (ABMA)
Website: www.theabma.org

Animal Behavior Society (ABS)
Website: http://animalbehaviorsociety.org
Facebook: www.facebook.com/group.php?gid=96833479864

American Biological Safety Association (ABSA)
Website: www.absa.org
Facebook: www.facebook.com/group.php?gid=115321105174916
Twitter: twitter.com/ABSAOffice

American College of Laboratory Animal Medicine (ACLAM)
Website: www.aclam.org
Email: mwbaclam@gsinet.net

American College of Veterinary Anesthesiologists (ACVA)
Website: www.acva.org

American College of Veterinary Pathologists (ACVP)
Website: www.acvp.org
Facebook: www.facebook.com/group.php?gid=130386390345733

American Committee on Laboratory Animal Diseases (ACLAD)
Website: www.aclad.org

American Council on Science and Health (ACSH)
Website: www.acsh.org
Facebook: www.facebook.com/pages/
    American-Council-on-Science-and-Health/110647352281294

Americans for Medical Progress (AMP)
Website: www.amprogress.org
Facebook: www.facebook.com/group.php?gid=30494293423
Twitter: twitter.com/medicalprogress

American Society of Laboratory Animal Practitioners (ASLAP)
PO Box 125
Adamstown, MD 21710
Telephone: 301-874-4826
Fax: 301-874-6195
Email: aslap-info@aslap.org
Website: www.aslap.org

American Physiological Society (APS)
Website: www.the-aps.org
Facebook: www.facebook.com/AmericanPhysiologicalSociety
Twitter: twitter.com/ExecDirectorAPS

American Veterinary Medical Association (AVMA)
Website: www.avma.org
Facebook: www.facebook.com/avmavets
Twitter: twitter.com/AVMAvets

Animal and Plant Health Inspection Service (APHIS)
Website: www.aphis.usda.gov
Twitter: twitter.com/APHISgov

Animals and Society Institute (ASI)
Website: www.animalsandsociety.org
Facebook: www.facebook.com/AnimalsandSocietyInstitute?v=
    app_2344061033
Twitter: twitter.com/onlineasi

Animal Transport Association (ATA)
Website: www.animaltransportationassociation.org/index.cfm

Animal Welfare Information Center (AWIC)
Website: awic.nal.usda.gov
Twitter: twitter.com/AnimalWelfareIC

Animal Welfare Institute (AWI)
Website: www.awionline.org
Facebook: www.facebook.com/animalwelfareinstitute
Twitter: twitter.com/AWIOnline

Armed Forces Institute of Pathology Department of Veterinary Pathology (AFIP)
Website: www.afip.org/consultation/vet path

Asian Federation for Laboratory Animal Science (AFLAS)
Website: www.aflas-office.org

Asian Mouse Mutagenesis Resource Association (AMMRA)
Website: www.ammra.info

Asia Pacific Biosafety Association (APBA)
Website: www.a-pba.org

Asociación Argentina de Ciencia y Tecnología de Animales de Laboratorio (AACyTAL)
Website: www.aacytal.com.ar/

Asociación Chilena de Ciencias del Animal de Laboratorio (ASOCHICAL)

Asociación Mexicana de la Ciencia de los Animales de Laboratorio (AMCAL)
Blog: amcal-ac.blogspot.com

Asociación Peruana de Ciencias del Animal de Laboratorio (APECAL)

Asociación Uruguaya de Ciencia y Tecnología de Animales de Laboratorio (AUCyTAL)
Website: www.aucytal.org

Asociación Venezolana para las Ciencias de Laboratorio (AVECAL)
Website: www.asovac.org

Association for the Assessment and Accreditation of Laboratory Animal Care International (AAALAC)
International Main Office
5283 Corporate Dr. Ste. 203
Frederick, MD 21703-2879 USA
Telephone: 301-696-9626
Fax: 301-696-9627
Email: accredit@aaalac.org
Website: aaalac.org

International European Office
Apartado de Correos 266
31080 Pamplona, Spain
Telephone: +34 948 100026
Fax: +34 948 100034
Email: euope@aaalac.org
Website: aaalac.org

International Pacific Rim Office
5283 Corporate Dr. Ste. 203
Frederick, MD 21703-2879 USA
Telephone: 301-696-9626
Fax: 301-696-9627
Email: pacificrim@aaalac.org
Website: aaalac.org

International Southeast Asia Office
61/370 Moo 3, Teparak Rd.
Bangpla, Bangplee
Samutprakarn, Thailand, 10540
Telephone: +662 175 5918
Email: seasia@aaalac.org
Website: aaalac.org

Association Française des Sciences et Techniques de l'Animal de Laboratoire (AFSTAL)
Website: www.afstal.com

Association for Gnotobiotics (AGS)
Website: www.gnotobiotics.org

Association of the British Pharmaceutical Industry (ABPI)
Website: www.abpi.org.uk
Facebook: www.facebook.com/pages/
  British-Pharmacological-Society/99860469636
Twitter: twitter.com/ABPI_UK

Association of Veterinary Anesthetists (AVA)
Website: www.ava.eu.com

Association Tunisienne des Sciences des Animaux de Laboratoire (ATSAL)
Email: atsal@topnet.tn

Associazione Italiana per le Scienze degli Animali da Laboratorio (AISAL)
Website: www.aisal.org

Australian and New Zealand Council for the Care of Animals in Research and Teaching
ANZCCART Australia
The University of Adelaide, SA 5005 Australia
Telephone: +61 8 8303 7585
Email: anzccart@adelaide.edu.au
Website: www.adelaide.edu.au/anzccart

ANZCCART New Zealand
c/o The Royal Society of New Zealand
PO Box 598 Wellington, New Zealand
Telephone: +64 4 472 7421
Email: anzccart@rsnz.org
Website: www.rsnz.org/advisory/anzccart

Australia and New Zealand Laboratory Animal Association (ANZLAA)
Website: www.anzlaa.org

Baltic Laboratory Animal Science Association (Balt-LASA)
Baltijos Laboratorinių Gyvūnų Mokslo Asociacija
Website: www.baltlasa.gf.vu.lt/lt/pradzia

Bangladesh Association for Laboratory Animal Science (BALAS)

Belgian Council for Laboratory Animal Science (BCLAS)
Website: bclas.be/index.php

BIOCOM
Website: biocom.org
Twitter: twitter.com/BIOCOMCA

Bioterios.com

British Veterinary Association (BVA)
Website: www.bva.co.uk

**British Veterinary Association Animal Welfare
Foundation (BVAAWF)**
Website: www.bva-awf.org.uk

California Biomedical Research Association (CBRA)
Website: ca-biomed.org

California Society for Biomedical Research (CSBR)
Website: ca-biomed.org/csbr/index.php

**Canadian Association for Laboratory Animal Medicine/
L'association Canadienne de la Médecine des Animaux de
Laboratoire (CALAM/ACMAL)**
Website: calam-acmal.org

**Canadian Association for Laboratory Animal Science/
Association Canadienne pour la Science des Animaux de
Laboratoire (CALAS/ACSAL)**
144 Front St. West, #640
Toronto, Canada M5J 2L7
Telephone: 416-593-0268
Fax: 416-979-1819
Email: office@calas-acsal.org
Website: www.calas-acsal.org

**Canadian Council on Animal Care/ Conseil Canadien de
Protection des Animaux (CCAC)**
Website: www.ccac.ca/en/CCAC_Main.htm

Center for Alternatives to Animal Testing (CAAT)
Website: caat.jhsph.edu
Facebook: www.facebook.com/JHUCAAT?ref=ts
Twitter: twitter.com/jhucaat

Centers for Disease Control and Prevention (CDC)
Website: www.cdc.gov
Facebook: www.facebook.com/CDC
Twitter: twitter.com/CDCgov

Central American, Caribbean and Mexican Association of Laboratory Animals (ACCMAL)

Centre for Documentation and Evaluation of Alternatives to Animal Experiments (ZEBET)
Website: www.bfr.bund.de/cd/1591

Chinese Academy of Sciences (SLACCAS)
Website: english.sibs.cas.cn/rs/fs/ShanghaiLaboratory
    AnimalCenterCAS/

Chinese Association for Laboratory Animal Science (CALAS)
Website: www.calas.org.cn

Chinese Society for Laboratory Animal Science (CSLAS, Taiwan)
Website: www.cslas.org

Colégio Brasileiro de Experimentação Animal (COBEA)
Website: www.cobea.org.br

Croatia Society for Laboratory Animals Science

Czech Laboratory Animal Science Association (CLASA)

Digital Resources for Veterinary Trainers (Digires)
Website: www.digires.co.uk

Eagleson Institute
Website: www.eagleson.org
Facebook: www.facebook.com/pages/
    Eagleson-Institute/91406671296

European Biomedical Research Association (EBRA)
Website: www.ebra.org

European Centre for the Validation of Alternative Methods (ECVAM)
Website: ecvam.jrc.ec.europa.eu

European College of Laboratory Animal Medicine (ECLAM)
Website: www.eclam.org

European College of Veterinary Anaesthesia and Analgesia (ECVAA)
Website: www.ecva.eu.com

European Consensus-Platform for Alternatives (ECOPA)
Website: www.ecopa.eu

European Federation of Animal Technologists (EFAT)
Website: www.efat.eu.com

European Federation of Pharmaceutical Industries and Associations (EFPIA)
Website: www.efpia.org
Twitter: twitter.com/EFPIA

European Resource Centre for Alternatives in Higher Education (EURCA)
Website: www.eurca.org

European Society for Laboratory Animal Veterinarians (ESLAV)
Website: www.eslav.org

European Society of Toxicology in Vitro (ESTIV)
Website: www.estiv.org

European Veterinarians in Education, Research and Industry (EVERI)
Website: www.fve.org/about_fve/sections/EVERI.html

Federation of American Societies for Experimental Biology (FASEB)
Website: www.faseb.org
Facebook: www.facebook.com/pages/Federation-of-American-
    Societies-for-Experimental-Biology/33038008112
Twitter: twitter.com/FASEBopa

Federation of European Laboratory Animal Science Associations (FELASA)
Website: www.felasa.eu

Federation of South American Societies and Associations of Laboratory Animal Science Specialists (FESSACAL)
[Argentina, Brazil, Chile, Colombia, Ecuador, Uruguay and Venezuela]

Federation of Veterinarians of Europe (FVE)
Website: www.fve.org

Finland Laboratory Animal Science (FinLAS)
Website: www.uku.fi/jarjestot/FinLAS

Fondazione Guido Bernardini (FGB)
Website: www.fondazioneguidobernardini.org

Food and Drug Administration (FDA)
Website: www.fda.gov
Facebook: www.facebook.com/FDA

Foundation for Biomedical Research (FBR)
Website: www.fbresearch.org
Facebook: www.facebook.com/FBResearch

Frozen Embryo and Sperm Archive (FESA)
Website: www.har.mrc.ac.uk/services/fesa

Fund for the Replacement of Animals in Medical Experiments (FRAME)
Website: www.frame.org.uk
Facebook: www.facebook.com/FRAMEpage?ref=nf
Twitter: twitter.com/FRAME_campaigns

Die Gesellschaft für Versuchstierkunde (GV SOLAS)
Website: www.gv-solas.de

Hellenic Society of Biomedical and Laboratory Animal Science (HSBLAS)
Website: hsblas.gr/index_en.php

Hungarian Laboratory Animal Scientists' Society (HLS)

Institute of Animal Technology (IAT)
Website: www.iat.org.uk

Institute of In Vitro Sciences (IIVS)
Website: www.iivs.org
Facebook: www.facebook.com/pages/
    The-Institute-for-In-Vitro-Sciences-Inc/153611251343473
Twitter: twitter.com/the_iivs/

Institute for Laboratory Animal Resources (ILAR)
500 Fifth St. NW
Washington, DC 20001 USA
Telephone: 202-334-2590
Fax: 202-334-1687
Website: dels.nas.edu/ilar

Institute of Laboratory Animal Science (ILAS)
Website: http://www.ltk.unizh.ch/de/dyn_output.html?content.
  vname=home_de

Interagency Coordinating Committee on the Validation of Alternative Methods (ICCVAM)
Website: iccvam.niehs.nih.gov

International Association of Colleges of Laboratory Animal Medicine (IACLAM)
Website: www.iaclam.org/index.html

International Committee on Taxonomy of Viruses (ICTV)
Website: www.ictvonline.org

International Council for Laboratory Animal Science (ICLAS)
Dr. Cecilia Carbone, Secretary General
CC 296 CP 1900, La Plata, Argentina
Email: ccarbone@fcv.unlp.edu.ar
Telephone: +54 221 421 1276
Fax: +54 221 421 1276
Website: iclas.org

International Federation of Biosafety Associations (IFBA)
Website: www.internationalbiosafety.org/english/index.asp

International Society for Applied Ethology (ISAE)
Website: www.applied-ethology.org

International Society for Transgenic Technology (ISTT)
Website: www.transtechsociety.org
Facebook: www.facebook.com/pages/International-Society-for-
  Transgenic-Technologies-ISTT/147397341246
Twitter: twitter.com/ISTT_TG

International Union of Basic and Clinical Pharmacology (IUPHAR)
Website: www.iuphar.org

International Union of Biological Sciences (IUBS)
Website: www.iubs.org

International Union of Immunological Societies (IUIS)
Website: www.iuisonline.org

International Union of Nutritional Sciences (IUNS)
Website: www.iuns.org

International Union of Physiological Sciences (IUPS)
Website: www.iups.org
Facebook: www.facebook.com/pages/International-Union-of-
     Physiological-Sciences-IUPS/184543411279

International Union of Toxicology (IUTOX)
Website: www.iutox.org

International Veterinary Academy of Pain Management (IVAPM)
Website: www.ivapm.org
Facebook: www.facebook.com/IVAPM.Page

In Vitro Testing Industrial Platform (IVTIP)
Website: www.ivtip.org

Israeli Laboratory Animal Forum (ILAF)
Website: www.ilaf.org.il

Japanese Association for Laboratory Animal Medicine (JALAM)
Website: http://plaza.umin.ac.jp/JALAM

Japanese Association for Laboratory Animal Science (JALAS)
Website: www.soc.nii.ac.jp/jalas/index_e.html

Japanese College of Laboratory Animal Medicine (JCLAM)
Website: http://plaza.umin.ac.jp/JALAM

Japanese Society for Laboratory Animal Resources (JSLAR)
Website: www.nichidokyo.or.jp

Japanese Society of Animal Models of Human Diseases (JSAMHD)
Website: http://133.1.15.131/ModelSoc/index.cfm

Japan Mouse/Rat Strain Resource Database (JMSR)
Website: www.shigen.nig.ac.jp/mouse/jmsr/top.jsp

Knock Out Rat Consortium (KORC)
Website: www.knockoutrat.org/index.php

Korean Association for Laboratory Animal Science (KALAS)
Website: www.kalas.or.kr

Korean College of Laboratory Animal Medicine (KCLAM)
Website: www.kclam.org

Korea Research Institute of Bioscience & Biotechnology (KRIBB)
Website: www.kribb.re.kr/eng

Kyoto University Rat Mutant Archive (KURMA)
Website: www.anim.med.kyoto-u.ac.jp/enu/archive.aspx

Laboratory Animal Breeders' Association (LABA)
Website: www.laba-uk.com

Laboratory Animal Management Association (LAMA)
7500 Flying Cloud Dr. Ste. 900
Eden Prairie, MN 55344 USA
Telephone: 952-253-6235
Fax: 952-835-4774
Website: www.lama-online.org

Laboratory Animal Science Association of Malaysia (LASAM)
Website: www.medic.ukm.my/laru/LASAM.htm

Laboratory Animal Science Association (United Kingdom) (LASA)
Website: www.lasa.co.uk

Laboratory Animal Scientist's Association in India (LASA)
Website: www.lasaindia.org

Laboratory Animals Veterinary Association (LAVA)
Website: www.lava.uk.net

Laboratory Animal Training Association (LATA)
Website: www.latanet.com

Laboratory Animal Welfare Training Exchange (LAWTE)
Website: www.lawte.org

Lhasa Limited
Website: www.lhasalimited.org

Massachusetts Society for Medical Research (MSMR)
Website: www.msmr.org

Michigan Society for Medical Research (MISMR)
Website: www.mismr.org

Mid-Continent Association for Agriculture, Biomedical Research and Education (MAABRE)
Website: www.maabre.info

National Association for Biomedical Research (NABR)
Website: www.nabr.org
Twitter: twitter.com/NABRorg

National BioResource Program for the Rat in Japan (NBRP)
Website: www.anim.med.kyoto-u.ac.jp/nbr/default.aspx

National Center for Laboratory Animal Science (NCLAS)
Website: www.ninindia.org/nclas.htm

National Centre for the Replacement, Refinement, and Reduction of Animals in Research (NC3Rs)
Website: www.nc3rs.org.uk

National Laboratory Animal Center (NLAC)
Finland Website: www.uku.fi/vkek
Taiwan Website: www.nlac.org.tw
Thailand Website: www.nlac.mahidol.ac.th/nlacmuEN

Nederlandse Vereniging voor Proefdierkunde/Biotechnische Vereniging (NVP/BV)
Website: www.proefdierkunde.nl

Netherlands Center for Alternatives to Animal Use (NCA)
Website: www.nca-nl.org

Non-animal Methods for Toxicity Testing (AltTox)
Website: www.alttox.org

Nordic Information Centre for Alternative Methods (NICA)
Website: www.cctoxconsulting.a.se/nica.htm

North Carolina Association for Biomedical Research (NCABR)
Website: www.ncabr.org
Facebook: www.facebook.com/NCABR

Northwest Association for Biomedical Research (NWABR)
Website: www.nwabr.org

Norwegian Reference Centre for Laboratory Animal Science and Alternatives (NORINA)
Website: oslovet.veths.no/fag.aspx?fag=57

Office International des Epizooties/World Organization for Animation Health (OIE)
Website: www.oie.int

Office of Laboratory Animal Welfare (OLAW)
Website: grants.nih.gov/grants/olaw/olaw.htm

Philippine Association for Laboratory Animal Science (PALAS)

Public Responsibility in Medicine and Research (PRIM&R)
Website: www.primr.org
Facebook: www.facebook.com/pages/
    Public-Responsibility-in-Medicine-and-Research/18688968156

Rat Resource and Research Center (RRRC)
Website: www.rrrc.us

Rat Strains (Rat Map)
Website: ratmap.gen.gu.se

Research Defence Society (RDS)
Website: www.understandinganimalresearch.org.uk
Facebook: www.facebook.com/UnderstandingAnimalResearch
Twitter: twitter.com/animalevidence

Royal College of Veterinary Surgeons (RCVS)
Website: www.rcvs.org.uk
Twitter: twitter.com/rcvs_uk

Scandinavian Laboratory Animal Science Association (SCAND-LAS)
Website: www.scandlas.org

Schweizerische Gesellschaft für Versuchstierkunde (SGV)
[Swiss Laboratory Animal Science Association]
Website: www.sgv.org

Scientists Center for Animal Welfare (SCAW)
Website: www.scaw.com

Shanghai Laboratory Animal Center, Chinese Academy of Sciences (SLACCAS)
Website: english.sibs.cas.cn/rs/fs/Shanghai
    LaboratoryAnimalCenterCAS/

Singapore Association for Laboratory Animal Science (SALAS)
Website: www.salas.sg

Sociedade Brasileira de Ciencia de Animais de Laboratorio (SBCAL)
Website: www.cobea.org.br/index.php

La Sociedad Cubana de la Ciencia del Animal de Laboratorio (SCCAL)

Sociedade Portuguesa de Ciências em Animais de Laboratório (SPCAL)
Website: www.spcal.pt/home

Sociedad Española de Experimentación Animal (SEEA)

Sociedad Española Para las Ciencias del Animal De Laboratorio (SECAL)
Website: www.secal.es

Society for in vitro Biology (SIVB)
Website: www.sivb.org

South African Association for Laboratory Animal Science (SAALAS)
Website: www.saalas.org

South African Society for Animal Science (SASAS)
Website: www.sasas.co.za

Thai Association for Laboratory Animal Science (TALAS)
Contact: Wantanee Ratanasak; DVM, MC
Email: vswrt@mahidol.ac.th

Turkey Laboratory Animal Science Association (TALAS, Turkey)

Universities Federation for Animal Welfare (UFAW)
Website: www.ufaw.org.uk

University of California Center for Animal Alternatives (UCCAA)
Website: www.vetmed.ucdavis.edu/Animal_Alternatives/main.htm

World Health Organization (WHO)
Website: www.who.int/en

World Organization for Animal Health (OIE)
Website: www.oie.int

publications

Periodicals

ALN (Animal Lab News) Magazine
  Website: www.labanimal.com/laban/index.html
  Website: www.alnmag.com

Animal Technology and Welfare, the official journal of the Institute of Animal Technology and of the European Federation of Animal Technologist *Animal Welfare Journal*, published by UFAW (see above).
  Website: www.ufaw.org.uk/animal.php

Applied Animal Behaviour Science, official journal of the International Society for Applied Ethology

Comparative Medicine, published 6 times a year by AALAS (see above)

Website: www.aalas.org

Emerging Infectious Disease

Website: www.cdc.gov/ncidod/EID/index.htm

Experimental Animals, the official journal of JALAS (see above)

Website: www.jstage.jst.go.jp/browse/expanim

ILAR Journal, published by ILAR (see above)

Journal of the American Association for Laboratory Animal Science, published by AALAS (see above)

Journal of Applied Animal Welfare Science, a joint project of Animals and Society Institute and the American Society for the Prevention of Cruelty to Animals

Journal of Visualized Experiments

Website: www.jove.com

Lab Animal, published by Nature Publishing Co.

Website: www.labanimal.com

Laboratory Animals, the official, quarterly journal of FELASA, GV-SOLAS, ILAF, LASA, NVP, SECAL, and SGV

Website: la.rsmjournals.com

Latin America Journal of Laboratory Animal Science

Website: www.bioterios.com/journal

Morbidity and Mortality Weekly Report

Website: www.cdc.gov/mmwr

Scandinavian Journal of Laboratory Animal Science, the official quarterly journal of the Scandinavian Society for Laboratory Animal Science

Website: biomedicum.ut.ee/sjlas/instructions.html

Textbooks

Select textbooks are listed below. The use of a particular vendor (e.g., Amazon) does not indicate a preference for that vendor. Texts marked with an asterisk (*) are available in electronic format (e.g., PDF, CD, Kindle, iPad).

Assistant Laboratory Animal Technician Training Manual. 2010, AALAS

Anatomy and Dissection of the Rat, 3rd edition. 1997, W. H. Freeman

*Animal Clinical Chemistry: A Practical Handbook for Toxicologists and Biomedical Researchers, 2nd edition. 2009, CRC Press

Anesthesia and Analgesia in Laboratory Animals, 2nd edition (ACLAM Series). 2008, Academic Press

*Animal Hematotoxicology: A Practical Guide for Toxicologists and Biomedical Researchers. 2008, CRC Press

Animal Models in Toxicology, 2nd edition. 2006, Informa Healthcare

Atlas of the Developing Rat Nervous System. 2008, Academic Press

Atlas of the Hematology of the Laboratory Rat. 1998, Elsevier Science

Atlas of the Neonatal Rat Brain. 2011, CRC Press

The Behavior of the Laboratory Rat: A Handbook with Tests. 2004, Oxford University Press, USA

Biodefense Research Methodology and Animal Models, 2nd edition, 2012, CRC Press

The Biology and Medicine of Rabbits and Rodents, 5th edition. 2010, Wiley-Blackwell

Chemoarchitectonic Atlas of the Rat Brain, 2nd edition. 2008, Academic Press

Loeb and Quimby's Clinical Chemistry of Laboratory Animals, 3rd edition (ACLAM Series). 2013, CRC Press

Clinical Laboratory Animal Medicine: An Introduction. 2006, Wiley-Blackwell

Exotic Companion Medicine Handbook for Veterinarians. 2006, Zoological Education Network

Exotic Companion Medicine Handbook for Veterinarians. 1996, Zoological Education Network

*Experimental and Surgical Techniques in the Rat. 1992, Academic Press

*Experimental Surgical Models in the Laboratory Rat. 2009, CRC Press

*Formulary for Laboratory Animals, 3rd edition. 2005, Wiley-Blackwell

Guidelines for the Care and Use of Mammals in Neuroscience and Behavioral Research. 2003, National Academies Press

Guidelines for the Humane Transportation of Research Animals. 2006, National Academies Press

Handbook of Laboratory Animal Science, 3rd edition. 2011, CRC Press

Immunodeficient Rodents: A Guide to Their Immunobiology, Husbandry, and Use. 1989, National Academies Press

Infectious Diseases of Mice and Rats. 1991, National Academies Press

Laboratory Animal Anaesthesia, 3rd edition. 2009, Academic Press

Laboratory Animal Medicine, 2nd edition (ACLAM Series). 2002, Academic Press

The Laboratory Rat, 2nd edition (ACLAM Series). 2005, Academic Press

The Laboratory Rat (Handbook of Experimental Animals Series). 2000, CRC Press

Laboratory Rat Procedural Techniques. 2010, CRC Press

Laboratory Animal Technician Training Manual. 2010, AALAS

Laboratory Animal Technologist Training Manual. 2010, AALAS

The Merck Veterinary Manual. 2010, Merck

Natural Pathogens in Laboratory Animals: Their Effects on Research. 2003, ASM Press

Necropsy Guide: Rodents and Rabbits. 1988, CRC Press

Nutrient Requirements of Laboratory Animals, 4th revised edition. 1995, National Academies Press

Occupational Health and Safety in the Care and Use of Research Animals. 1997, National Academies Press

Pain Management in Animals. 2000, Saunders

Pathology of Laboratory Rodents and Rabbits, 3rd edition. 2007, Wiley-Blackwell

Flynn's Parasites of Laboratory Animals, 2nd edition (ACLAM Series). 2007, Wiley-Blackwell

Planning and Designing Research Animal Facilities (ACLAM Series). 2008, Academic Press

Principles of Laboratory Animal Science: A Contribution to the Humane Care of Animals and to the Quality of Experimental Results, revised edition. 2001, Elsevier

The Rat Brain in Stereotaxic Coordinates, 6th edition. 2007, Academic Press

*Rat Jugular Vain and Carotid Artery Catheterization for Acute Survival Studies: A Practical Guide. 2006, Springer

The Rat Nervous System, 3rd edition. 2004, Academic Press

*Recognition and Alleviation of Pain in Laboratory Animals. 2009, National Academies Press

*Recognition and Alleviation of Distress in Laboratory Animals. 2008, National Academies Press

Research Techniques in the Rat. 1982, Charles C. Thomas Publisher Ltd

*Rodents. 1996, National Academies Press

Stereotaxic Surgery in the Rat: A Photographic Series. 2005, A. J. Kirby Co.

*Strategies That Influence Cost Containment in Animal Research Facilities. 2000, National Academies Press

The UFAW Handbook on the Care and Management of Laboratory Animals, 8th edition. 2010, Wiley-Blackwell

Using Animal Models in Biomedical Research: A Primer for the Investigator. 2008, World Scientific Publishing Company

---

electronic resources

Websites

ALTBIB
Bibliography on the Alternatives to Animal Testing
Website: toxnet.nlm.nih.gov/altbib.html

Animalresearch.info
Website: www.animalresearch.info/en/home

BioMed Search
Website: www.biomedsearch.com

Biotech Gate
Website: www.biotechgate.com/gate/db/index.php

Biotechnology Information Institute
Website: www.bioinfo.com

Charles Louis Davis Foundation for the Advancement of Veterinary Pathology
Website: www.cldavis.org

Cold Spring Harbor Laboratory
Website: www.cshl.edu

Communicable Disease Surveillance Centre (UK)
Website: www.publichealth.hscni.net

Embryology—Rat Development
Website: embryology.med.unsw.edu.au/OtherEmb/Rat.htm

Enrichment Record
Website: enrichmentrecord.com

EURATRANS
European Large-scale Functional Genomics in the Rat for
    Translational Research
Website: www.euratrans.eu

Federal Register
Website: www.federalregister.gov

Food and Drug Administration Animal & Veterinary
Website: www.fda.gov/AnimalVeterinary/default.htm

Hamner Institutes for Health Sciences (formerly CIIT)
Website: www.thehamner.org

International Laboratory Code Registry
Website: dels.nas.edu/global/ilar/Lab-Codes

Jackson Laboratory Bioinformatics
Website: www.informatics.jax.org

Journal of Visualized Experiments
Website: www.jove.com

Laboratory of Neuro Imaging—Rat Atlas
Website: www.loni.ucla.edu/Atlases/Atlas_Detail.jsp?atlas_id=1

Medical, Clinical, and Occupational Toxicology
Resource Home Page
Website: www.pslgroup.com/dg/39fe.htm

Medical Education Information Center
Website: www.uth.tmc.edu/pathology/medic/index.html

Medical Matrix—Guide to Internet Clinical Medicine Resources
Website: http://www.medmatrix.org

Medical Research Council (UK)
Website: www.mrc.ac.uk/index.htm

Medscape
Website: www.medscape.com

Metris
Website: www.metris.nl

Microinjection Workshop
Website: www.microinjectionworkshop.net

National Academy of Science
Website: www.nas.edu

National Agricultural Library (US)
Website: www.nal.usda.gov

National BioResource Project for the Rat in Japan
Website: www.anim.med.kyoto-u.ac.jp/nbr

National Cancer Institute
Website: www.cancer.gov

National Eye Institute
Website: www.nei.nih.gov

National Health Information Center
Website: www.health.gov/nhic

National Heart, Lung, and Blood Institute
Website: www.nhlbi.nih.gov

National Human Genome Research Institute
Website: www.genome.gov

National Institute of Allergy and Infectious Diseases
Website: www.niaid.nih.gov

National Institute of Arthritis and Musculoskeletal and Skin Diseases
Website: www.niams.nih.gov

National Institute of Biomedical Imaging and Bioengineering
Website: www.nibib.nih.gov

National Institute of Child Health and Human Development
Website: www.nichd.nih.gov

National Institute of Dental and Craniofacial Research
Website: www.nidcr.nih.gov

National Institute of Diabetes and Digestive and Kidney Diseases
Website: www2.niddk.nih.gov

National Institute of Environmental Health Sciences
Website: www.niehs.nih.gov

National Institute of Mental Health
Website: www.nimh.nih.gov

National Institute of Neurological Disorders and Stroke
Website: www.ninds.nih.gov

National Institute of Nursing Research
Website: www.ninr.nih.gov

National Institute on Aging
Website: www.nia.nih.gov

National Institute on Alcohol Abuse and Alcoholism
Website: www.niaaa.nih.gov

National Institute on Drug Abuse
Website: www.drugabuse.gov

National Institute on Deafness and Other Communication Disorders
Website: www.nidcd.nih.gov

National Institute on Minority Health and Health Disparities
Website: www.nimhd.nih.gov

National Institutes of Health
Website: www.nih.gov

National Institutes of Health Library
Website: nihlibrary.nih.gov

National Library of Medicine (US)
Website: www.nlm.nih.gov

Nature: The Rat Genome
Website: www.nature.com/nature/focus/ratgenome

Net Vet/Electronic Zoo
Website: netvet.wustl.edu

Pan American Health Organization
Website: www.paho.org

Peromyscus Genetic Stock Center
Website: stkctr.biol.sc.edu

Pro-Test
Website: www.pro-test.org.uk

Rat ENU Mutagenesis
Website: www.anim.med.kyoto-u.ac.jp/enu

Rat Genome Database
Website: rgd.mcw.edu

Rat Genome Project (Baylor College of Medicine)
Website: www.hgsc.bcm.tmc.edu/project-species-m-Rat.
  hgsc?pageLocation=Rat

Rat Genome Resources
Website: www.ncbi.nlm.nih.gov/genome/guide/rat/index.html

Rat Mapping Resources
Website: www.biologydir.com/rat-mapping-resources-
    info-5377.html

Rat Resource and Research Center
Website: www.rrrc.us

Rat Strain Index
Website: www.informatics.jax.org/external/festing/rat/STRAINS.
    shtml

Rat Strains/Rat Map
Website: ratmap.gen.gu.se

Retinal Degeneration Rat Model Resource
Website: www.ucsfeye.net/mlavailRDratmodels.shtml

Science NetLinks
Website: www.sciencenetlinks.com

Understanding Animal Research
Website: www.understandinganimalresearch.org.uk

**University of California Davis Center for Animal Alternatives
Information**
Website: www.lib.ucdavis.edu/dept/animal alternatives

U.S. Government Printing Office
Website: www.gpo.gov

Veterinary Bioscience Institute
Website: vetbiotech.com

VetTech/Vet Tech School Guide
Website: www.vettech.org

Wistar Institute
Website: www.wistar.org

Listserv Mailing Lists

CompMed
Listserv@Listserv.AALAS.org

Human-Animal Studies Listserv
Website: www.animalsandsociety.org/pages/
    human-animal-studies-listserv

IACUC-Forum
Website (Application Form): www.iacuc.org/IACUC_FORUM_
    application.pdf

LAWTE Listserv
Listserv@listserv.aalas.org

Office of Laboratory Animal Welfare (OLAW) Listserv
Listserv@list.nih.gov

ProMed AHEAD
Website: www.isid.org/promedmail/subscribe.lasso

PsycEXTRA Listserv
Website: www.apa.org/pubs/databases/psycextra/archive.aspx

Research Animals Topics Discussion Forum
RAT-TALK@NIC.Surfnet.NL

TechLink
Listserv@Listserv.AALAS.org

Select Databases

BioMed Central
Website: www.biomedcentral.com

BioMedical Engineering Network
Website: engineering.purdue.edu/BMEnet

Cochrane Collaboration
Website: www.cochrane.org

Chemical Carcinogenesis Research Information System (CCRIS)
Website: toxnet.nlm.nih.gov/cgi-bin/sis/htmlgen?CCRIS

Genetic Toxicology Gene Bank (Gene-Tox)
Website: toxnet.nlm.nih.gov/cgi-bin/sis/htmlgen?GENETOX

Integrated Risk Information System
Website: www.epa.gov/IRIS

Israel Science and Technology Homepage: Biomedical Databases
Website: www.science.co.il/biomedical-databases.asp

PubMed
Website: www.pubmed.gov

Rat Genome Database
Website: rgd.mcw.edu

Toxnet
Toxicology Data Network
Website: www.toxnet.nlm.nih.gov

equipment and facilities meetings

AALAS National Meeting
Website: nationalmeeting.
aalas.org

TurnKey Conference
Website: www.
turnkeyconference.com

Tradeline Conference
Website: www.tradelineinc.com

technical education

Master of Laboratory Animal Science (MLAS)
Office of Professional Studies in the Health Sciences
Drexel University College of Medicine
245 North 15th St., Mail Stop 344, Room 4104 NCB
Philadelphia, PA 19102 USA

Telephone: 215-762-4692
Fax: 215-762-7434
Email: medicalsciences@drexelmed.edu
Website: http://drexel.edu/Home/AcademicPrograms/
  ProfessionalStudiesintheHealthSciences/
  AnimalSciencePrograms/
  MasterofLaboratoryAnimalScienceMLASProgram.aspx

Purdue University
Veterinary Technology
School of Veterinary Medicine
Distance Learning Program
625 Harrison St.
West Lafayette, IN 47907 USA
Telephone: 765-496-6579
Email: vettech@purdue.edu
Website: www.vet.purdue.edu/vettech

vendors

Animals

Ace Animals, Inc.
PO Box 122
Boyertown, PA 19512 USA
Telephone: 610-367-6047
Fax: 610-367-6048
Email: info@aceanimals.com
Web: www.aceanimals.com

Animal Resources Centre
PO Box 1180
Canning Vale WA 6970 Australia
Telephone: (+61 8) 9332 5033
Fax: (+61 8) 9310 2839
Email: orders@arc.wa.gov.au
Web: www.arc.wa.gov.au

B&K Universal, Ltd.
Grimston, Aldbrough
Hull, East Yorkshire HU11
  4QE UK
Telephone: +44 1964 527555
Fax: +44 1964 527006
Email: salesbku@aol.com
Web: www.bku.com

BioLASCO Taiwan Co., Ltd.
3F, 316, Chung-Yang Rd.
Taipei 115 Taiwan
Telephone: 886-2-2651-8669
Fax: 886-2-2651-8969
Email: lasco@biolasco.com.tw
Web: www.biolasco.com.tw

Centre for Animal Resources (CARe)
28 Medical Dr.
Centre for Life Sciences #05-02
National University of Singapore
Singapore 117456
Telephone: (+65) 6791-3039
Fax: (+65) 6794-4058
Email: lacsec@nus.edu.sg
Web: www.nus.edu.sg/lac

Charles River Laboratories
251 Ballardvale St.
Wilmington, MA 01887 USA
Toll Free: 800-522-7287
Telephone: 978-658-6000
Fax: 978-988-9236
Email: comments@crl.com
Web: www.criver.com

Harlan Laboratories, Inc.
8520 Allison Pointe Blvd.,
   Ste. 400
Indianapolis, IN 46250 USA
Telephone: 317-806-6080
Fax: 317-806-6068
Toll Free: 800-793-7287
Email: harlan@harlan.com
Web: www.harlan.com

Hilltop Lab Animals
PO Box 183, 131 Hilltop Dr.
Scottdale, PA 15683 USA
Telephone: 724-887-8480
Fax: 724-887-3582
Toll Free: 800-245-6921
Email: clientserve@hilltoplabs.
   com
Web: hilltoplabs.com

Ozgene Pty, Ltd.
PO Box 1128
Bentley DC, WA 6983 Australia
Telephone: +61 8 9212 2200
Fax: +61 8 9212 2299
Email: ozgene@ozgene.com
Web: www.ozgene.com

SAGE® Labs
PO Box 122
Boyertown, PA 19512 USA
Telephone: 610-367-6047
Fax: 610-367-6048
Email: info@
   sageresearchmodels.com
Web: www.sageresearchmodels.
   com

Scanbur
Box 6023, Vibyvagen 72
S192 06 Sollentuna, Sweden
Telephone: +46 8 754 00 65
Fax: +46 8 594 6767 97
Email: info-se@scanbur.eu
Web: www.nova-scb.com

Sierra for Medical Science, Inc.
PO Box 5692
Whittier, CA 90607 USA
Telephone: 562-695-5650
Fax: 562-695-5488
Email: info@sierra-medical.com
Web: www.sierra-medical.com

Simonsen Laboratories, Inc.
1180-C Day Rd.
Gilroy, CA 95020 USA
Telephone: 408-847-2002
Fax: 408-847-4176
Email: orders@simlab.com
Web: www.simlab.com

Taconic
One Hudson City Centre
Hudson, NY 12534 USA
Toll Free: 888-822-6642
Telephone: 518-697-3900
Fax: 518-697-3910
Email: custserv@taconic.com
Web: www.taconic.com

Bedding

Absorption Corp.
6960 Salashan Pkwy.
Ferndale, WA 98248 USA
Toll-free: 800-242-2287
Telephone: 360-734-7415
Fax: 360-671-1588
Email: absorbs@absorption-
corp.com
Web: www.absorptioncorp.com

Andersons/Bed-O'Cobs
PO Box 119
Maumee, OH 43537 USA
Toll Free: 866-234-0505,
ext. 6325
Telephone: 419-891-6325
Fax: 419-891-6539
Email: dr_cobs@andersonsinc.
com
Web: www.bedocobs.com

B & C Pulaski Corp. Ltd.
Bangkok, Thailand
323 Betagro Tower
North Park Viphavadee Rangsit
Road
Bangkok 10210 Thailand
Telephone: +66 2833-8683
Fax : +66 2833-8634
Email: wanneet@betagro.com
Email: voradej@betagro.com

Green Products Co.
PO Box 756
Conrad, IA 50621 USA
Toll Free: 800-247-7807
Telephone: 641-366-2001
Fax: 641-366-2366
E-Mail: gregg@greenproducts.
com
Web: www.greenproducts.com

Harlan Laboratories, Inc.
8520 Allison Pointe Blvd.,
Ste. 400
Indianapolis, IN 46250 USA
Toll Free: 800-793-7287
Telephone: 317-806-6080
Fax: 317-806-6068
E-Mail: harlan@harlan.com
Web: www.harlan.com

J Rettenmaier & Söhne
Holzmühle 1
Rosenberg, 73494 Germany
Telephone: +49 7967 152 153
Fax: +49 7967 152 500 153
E-Mail: ralf.hertl@jrs.de
Web: www.jrs.de

Ketchum Manufacturing, Inc.
1245 California Ave.
Brockville, ON K6V 7N5 Canada
Telephone: 613-342-8455
Fax: 613-342-7550
Email: ketchum@sympatico.ca
Web: www.ketchum.ca

Omni BioResources, Inc.
PO Box 8224
Cherry Hill, NJ 08002 USA
Telephone: 856-667-0168
Fax: 856-667-0124
Email: kcbecksr@aol.com

PJ Murphy Forest Products Corp.
PO Box 300
Montville, NJ 07045 USA
Telephone: 973-316-0800
Fax: 973-316-9455
Email: sales@pjmurphy.net
Web: www.pjmurphy.net

Shepherd Specialty Papers
PO Box 64
Watertown, TN 37184 USA
Toll Free: 800-882-5001
Fax: 615-237-3267
Email: jshepherd3@ssponline.com
Web: www.ssponline.com

Sivahkorn Trading Partnership
Siam Thornee Rd.
Bangkok, 10220 Thailand
Telephone: +668 409 38766
Email: luksikarnt@yahoo.co.th
Web: www.alibaba.com/
member/th101968673.html

Tapvei Oy
Vaikkojoentie 33
Kortteinen, 73620 Finland
Telephone: +358 17 581 0966
Fax: +358 17 581 0967
Email: info@tapvei.com

Caging and Equipment

Able Scientific
24 Sorbonne Crescent
Canningvale, Western Australia
6155
Australia
Telephone: +61 89456 2800
Fax: +61 89456 3800
Email: info@ablescientific.com.au
Web: www.ablescientific.com.au

Allentown, Inc.
PO Box 698
165 Rte. 526
Allentown, NJ 08501 USA
Toll Free: 800-762-2243
Telephone: 609-259-7951
Fax: 609-259-0449
Email: info@allentowninc.com
Web: www.allentowninc.com
Facebook: www.facebook.com/
allentowninc

**Alternative Design Mfg &
Supply, Inc.**
PO Box 6330
3055 Cheri Whitlock Dr.
Siloam Springs, AR 72761 USA
Toll Free: 800-320-2459
Telephone: 479-524-4343
Fax: 479-524-4125
Email: mail@altdesign.com
Web: www.altdesign.com

Ancare Corp.
PO Box 814
2647 Grand Ave.
Bellmore, NY 11710 USA
Toll Free: 800-645-6379
Telephone: 516-781-0755
Fax: 516-781-4937
Email: information@ancare.com
Web: www.ancare.com

Animal Care Systems, Inc.
7086 S. Revere Pkwy., Ste. 100
Centennial, CO 80112 USA
Toll Free: 888-827-3861
Telephone: 720-283-0177
Fax: 720-283-0179
Email: info@acs-dvm.com
Web: www.animalcaresystems.
  com

Arrowmight Biosciences
Rotherwas Industrial Estate
Campwood Rd.
Hereford, HR2 6JD UK
Telephone: +44 1432 379111
Fax: +44 1432 344960
Email: sales@arrowmight.com
Web: www.arrowmight.com

Aston Pharma, Ltd.
PO Box 225
Ely, Cambs CB7 9BB UK
Telephone: 05602 394 693
Email: info@aston-pharma.com
Web: www.aston-pharma.com

B.I.K. Industries
Khetwadi 5th Ln., Kensara
  House
Mumbai, Maharashtra 400 004
  India
Telephone: +91 22 23820722
Fax: +91 22 23808732
Email: info@bikindia.com
Web: www.bikindia.com

Bellflower Animal Care
825 Bellflower Ln.
Bolingbrook, IL 60440 USA
Telephone: 630-771-9887

bioBubble, Inc.
3024 W. Prospect Rd.
Fort Collins, CO 80526 USA
Telephone: 970-224-4262
Fax: 970-224-2419
Email: sales@biobubble.com
Web: www.biobubble.com

BioLASCO Taiwan Co., Ltd.
3F, 316, Chung-Yang.
Taipei 115 Taiwan
Telephone: 886-2-2651-8669
Fax: 886-2-2651-8969
Email: lasco@biolasco.com.tw
Web: www.biolasco.com.tw

Britz & Company
1302 9th St.
Wheatland, WY 82201 USA
Telephone: 307-322-4040
Fax: 307-322-4141
Web: www.britzco.com

Charles River Laboratories
251 Ballardvale St.
Wilmington, MA 01887 USA
Toll Free: 800-522-7287
Telephone: 978-658-6000
Fax: 978-988-9236
Email: comments@crl.com
Web: www.criver.com

Circle K Industries
25563 N. Gilmer Rd.
Mundelein, IL 60060 USA
Toll Free: 800-284-4005
Telephone: 847-949-0363
Fax: 847-566-7309
Email: sales@circle-k.com
Web: www.circle-k.com

Class Biologically Clean, Ltd. (CBC)
3202 Watford Way
Madison, WI 53713-4624 USA
Telephone: 608-273-9661
Fax: 608-273-9668
Email: info@cbclean.com
Web: www.cbclean.com

CliniPath, Ltd.
PO Box 523
Hull, H09 9HD UK
Telephone: +44 (0) 755 495 7336
Fax: +44 (0) 1482 213200
Email: sales@
  clinipathequipment.com
Web: www.clinipathequipment.
  com

E. Becker & Co. GmbH
Hermannstr. 2-8
44579 Castrop-Rauxel,
  Germany
Telephone: +49 2305 973040
Fax: +49 2305 9730-444
Email: www.info@ebeco.de
Web: www.ebeco-vth.de

EquipNet, Inc.
50 Hudson Rd.
Canton, MA 02021 USA
Telephone: 781-821-3482
Fax: 617-671-1269
Email: sales@equipnet.com
Web: www.equipnet.com

Ets. Scholz
Kasteelstraat 72
Overijse, 3090 Belgium
Telephone: +32 02 687 84 84
Fax: +32 02 687 86 38
Email: scholz.kuyt@pandora.be

Fabrications Pajon
2 Bis rue des Pins
Fleury les Aubrais, Loiret 45400
  France
Telephone: +33 23880 6197
Fax: +33 23883 9977
Email: pajonfab@fabrications-
  pajon.com
Web: www.fabrications-pajon.
  com

Genestil
1 rue du Mesnil
Royaucourt, 60420 France
Telephone: +33 3 44 51 73 88
Fax: +33 3 44 51 13 61
Email: info@genestil.com
Web: www.genestil.com

indulab ag
Haagerstr. 59
Gams, 9473 Switzerland
Telephone: +41 081 750 31 40
Fax: +41 081 750 31 45
Email: infos@indulab.ch
Web: www.indulab.ch

Innovive, Inc.
9689 Towne Centre Dr.
San Diego, CA 92121 USA
Toll Free: 866-43-CAGES
Telephone: 858-309-6620
Fax: 858-309-6621
Email: info@innoviveinc.com
Web: www.innoviveinc.com

IntelliBio
25 rue de Fossieux
Nomeny, 54610 France
Telephone: +33 3 83 28 16 04
Fax: +33 3 83 90 11 58
Email: info@intelli-bio.com
Web: www.intelli-bio.com

Lab Products, Inc.
742 Sussex Ave.
Seaford, DE 19973 USA
Toll Free: 800-526-0469
Telephone: 302-628-4300
Fax: 302-628-4309
Email: info@labproductsinc.com
Web: www.labproductsinc.com

LABeX of MA
100 Grove St.
Worcester, MA 01605 USA
Telephone: 508-755-2243
Fax: 508-755-2249
Email: edrusso@labexofma.com
Web: www.labexofma.com

Lenderking Caging Products
8370 Jumpers Hole Rd.
Millersville, MD 21108 USA
Telephone: 410-544-8795
Fax: 410-544-5069
Email: sales@lenderking.com
Web: www.lenderking.com

**LGL Animal Care
Products, Inc.**
721 Peach Creek Cut-Off Rd.
College Station, TX 77845 USA
Telephone: 979-690-3434
Fax: 979-690-8863
Email: inform@lglacp.com
Web: www.lglacp.com

Lignifer KFT
Ady Endre U47
Isaszeg, 2117 Hungary
Telephone: +36 28 495709/
    +36 28 495719
Fax: +36 28 494608
Email: lignifer@vnet.hu
Web: www.lignifer.hu

Lithgow Laboratory Services
8414 W. Farm Rd., Ste. 180-225
Las Vegas, NV 89131 USA
Telephone: 702-413-0832
Fax: 702-363-4981
Email: sales@lithgowservices.com
Web: www.lithgowservices.com

Maryland Plastics, Inc.
251 E. Central Ave.
Federalsburg, MD 21632 USA
Toll Free: 800-544-5582
Telephone: 410-754-5566
Fax: 410-754-8882
Email: sales@marylandplastics.
    com
Web: www.marylandplastics.com

Mini Mitter, A Respironics Company
20300 Empire Ave., Bldg. B-3
Bend, OR 97701 USA
Toll Free: 800-685-2999
Telephone: 541-598-3800
Fax: 541-322-7277
Email: mm@respironics.com
Web: www.minimitter.com

Momo Line SRL
Via Circunvallazione Esterna 12
Casandrino Naples, 80025 Italy
Telephone: +3981 505 3714
Fax: +3981 505 2783
Email: info@momoline.it
Web: www.momoline.it

North Kent Plastic Cages
Unit 4 Gills Ct., Chaucer Cl.
Medway City Estate
Rochester, Kent MEZ 4NR UK
Telephone: +44 01634 295888
Fax: +44 01634 725877
Email: info@northkentplastics.
  co.uk
Web: www.northkentplastics.
  co.uk

Otto Environmental, LLC
6914 N. 124th St.
Milwaukee, WI 53224 USA
Telephone: 414-358-1001
Fax: 414-358-9035
Email: sales@ottoenvironmental.
  com
Web: www.ottoenvironmental.
  com

Pacific Sentry, Inc.
7126 180th Ave. NE, Ste. C-106
Redmond, WA 98052 USA
Telephone: 425-497-8494
Fax: 425-497-8494
Email: sentry@photonicsystems.
  com
Web: www.pacificsentry.com/
  smallanimal.html

Park Bioservices, LLC
154 Center St.
Groveland, MA 01834 USA
Toll Free: 800-947-5226
Telephone: 978-794-8500
Fax: 978-794-0307
Email: info@parkbio.com
Web: www.parkbio.com

Plas-Labs, Inc.
401 E. North St.
Lansing, MI 48906 USA
Toll Free: 800-866-7527
Telephone: 517-372-7177
Fax: 517-372-2857
Email: sales@plas-labs.com
Web: www.plas-labs.com

Plexx B.V.
PO Box 86
Elst, 6660 AB The Netherlands
Telephone: +31 481 377 797
Fax: +31 481 377 910
Email: info@plexx.eu
Web: www.plexx.eu

Scanbur A/S
Silovej 16-18
Karlslunde, 2690 Denmark
Telephone: +45 5686 5600
Fax: +45 5682 1405
Email: info@scanbur.eu
Web: www.scanbur.eu

Suburban Surgical Co., Inc.
275 12th St.
Wheeling, IL 60090 USA
Toll Free: 800-323-7366
Telephone: 847-537-9320
Fax: 847-537-9061
Email: info@suburbansurgical.
   com
Web: www.suburbansurgical.com

Tecniplast S.P.A.
Via 1 Maggio, 6
Buguggiate, Varese 21020 Italy
Telephone: +39 0332 809711
Fax: +39 0332 458315
Email: tecnicom@tecniplast.it
Web: www.tecniplast.it

Three Shine, Inc.
1326 Guanpyung-Dong
Yuseong-Gu, Daejeon 305-509
Republic of Korea
Telephone: +82 42 933 3361
Toll Free: +82 42 933 3398
Email: SK@threeshine.com
Web: threeshine.com

Unifab Corp.
3030 Kersten Ct.
Kalamazoo, MI 49048 USA
Toll Free: 800-648-9569
Telephone: 269-382-2803
Fax: 269-382-2825
Email: jeff.kokmeyer@
   unifabcorp.com
Web: www.unifabcages.com

Uno Roestvaststaal b.v.
PO Box 15
Zevenaar, 6900AA The
   Netherlands
Telephone: +31 316 524451
Fax: +31 316 523785
Email: info@unobv.com

Diet

Altromin GmbH
Im Seelenkamp 20
Lage, NRW 32791 Germany
Telephone: +49 5232 60880
Fax: +49 5232 608820
Email: info@altromin.de
Web: www.altromin.de

Aston Pharma, Ltd
PO Box 225
Ely, Cambs CB7 9BB UK
Telephone: 05602 394 693
Email: info@aston-pharma.com
Web: www.aston-pharma.com

B&K Universal, Ltd.
A division of Marshall
   BioResources
Grimston, Aldbrough
Hull, East Yorkshire HU11 4QE
   UK
Telephone: +44 1964 527555
Fax: +44 1964 527006
Email: salesbku@aol.com
Web: www.bku.com

Bellflower Animal Care
825 Bellflower Ln.
Bolingbrook, IL 60440 USA
Telephone: 630-771-9887

Bio-Serv
One 8th St.
Frenchtown, NJ 08825 USA
Toll Free: 800-996-9908
Telephone: 908-996-2155
Fax: 908-996-4123
Email: sales@bio-serv.com
Web: www.bio-serv.com

ClearH2O, Inc.
117 Preble St.
Portland, ME 04101 USA
Telephone: 207-221-0039
Fax: 207-774-3449
Email: info@clearh2o.com
Web: www.clearh2o.com

Dyets, Inc.
2508 Easton Ave.
Bethlehem, PA 18017 USA
Toll Free: 800-275-3938
Telephone: 610-868-7701
Fax: 800-329-3938
Email: dyets@verizon.net
Web: www.dyets.com

Genestil
1 rue du Mesnil
Royaucourt, 60420 France
Telephone: +33 3 44 51 73 88
Fax: +33 3 44 51 13 61
Email: info@genestil.com
Web: www.genestil.com

Harlan Laboratories, Inc.
8520 Allison Pointe Blvd.,
   Ste. 400
Indianapolis, IN 46250 USA
Toll Free: 800-793-7287
Telephone: 317-806-6080
Fax: 317-806-6068
Email: harlan@harlan.com
Web: www.harlan.com

Lantmännen
Lidköping, 53187 Sweden
Telephone: +46 510 88664
Fax: +46 510 88600
Email: michael.maslov@
   lantmannen.com

LASvendi
Pagenstr. 68
Soest, 59494 Germany
Telephone: +49 2921 3456 890
Fax: +49 2921 3456 891
Email: info@lasvendi.com
Web: www.lasvendi.com

Lillico Biotechnology
PO Box 431
Hookwood, Surrey RH6
   0VW UK
Telephone: +44 1293 827940
Fax: +44 1293 782235
Email: saleslillico@aol.com

Mucedola
Via G. Galilei 4
Settimo Milanese, MI 20019
   Italy
Telephone: +39 02 489 155 81
Fax: +39 02 489 15695
Email: mucedola@mucedola.it
Web: www.mucedola.it

NLS Animal Health
11445 Cronridge Dr.
Owings Mills, MD 21117 USA
Toll Free: 800-638-8672
Telephone: 410-581-1800
Fax: 888-568-2825
Email: skirsh@nlsanimalhealth.
   com
Web: www.nlsanimalhealth.com

Provimi Kliba AG
Rinaustrasse
Kaiseraugst, 4303 Switzerland
Telephone: +41 61 816 16 16
Fax: +41 61 816 18 00
Email: kliba-nafag@provimi-
   kliba.ch
Web: www.kliba-nafag.ch

Purina LabDiet
PO Box 66812
St. Louis, MO 63144 USA
Toll Free: 800-227-8941
Telephone: 636-742-6295
Fax: 636-742-6166
Email: info@labdiet.com
Web: www.labdiet.com

Research Diets, Inc.
20 Jules Ln.
New Brunswick, NJ 08901 USA
Toll Free: 877-486-2486
Telephone: 732-247-2390
Fax: 732-247-2340
Email: info@researchdiets.com
Web: www.researchdiets.com

SAFE
Rte. de Saint-Bris
Augy, 89290 France
Telephone: +33 386 53 76 90
Fax: +33 386 53 35 96
Email: info@safe.evis.net
Web: www.safe-diets.com

Scanbur
Box 6023, Vibyvagen 72
S-192 6 Sollentuna, Sweden
Telephone: +46 8 594 6767 80
Fax: +46 8 594 6767 97
Email: orders-se@scanbur.eu
Web: www.nova-scb.com

Special Diets Services
PO Box 705
Witham, Essex CM8 3AD UK
Telephone: +44 1376 511260
Fax: +44 1376 511247
Email: info@sdsdiets.com
Web: www.sdsdiets.com

ssniff Spezialdiaeten GmbH
Ferdinand-Gabriel-Weg 16
Soest, 59494 Germany
Telephone: +49 292 196580
Fax: +49 292 1965840
Email: mail@ssniff.de

Zeigler Bros., Inc.
400 Gardners Station Rd.
Gardners, PA 17324 USA
Toll Free: 800-841-6800
Telephone: 717-677-6181
Fax: 717-677-6826
Email: sales@zeiglerfeed.com
Web: www.zeiglerfeed.com

Environmental Enrichment

Bio-Serv
One 8th St.
Frenchtown, NJ 08825 USA
Toll Free: 800-996-9908
Telephone: 908-996-2155
Fax: 908-996-4123
Email: sales@bio-serv.com
Web: www.bio-serv.com

Ketchum Manufacturing, Inc.
1245 California Ave.
Brockville, ON K6V 7N5 Canada
Telephone: 613-342-8455
Fax: 613-342-7550
Email: ketchum@sympatico.ca
Web: www.ketchum.ca

Lab Etc, Inc.
PO Box 111
Clayton, DE 19938 USA
Toll Free: 877-452-2382
Fax: 302-338-8822
Email: fish@labetc.net
Web: www.labetc.net

Lillico Biotechnology
PO Box 431
Surrey, RH6 0UW UK
Telephone: +44 (1293) 827-940
Fax: +44 (1293) 782-235
Email: saleslillico@aol.com
Web: www.lillicobiotech.co.uk

Lomir Biomedical, Inc.
95 Huot
Notre Dame De l'iIe
Perrot, QC J7V 7M4 Canada
Telephone: 514-425-3604
Fax: 514-425-3605
Email: info@lomir.com
Web: www.lomir.com

Otto Environmental, LLC
6914 N. 124th St.
Milwaukee, WI 53224 USA
Telephone: 414-358-1001
Fax: 414-358-9035
Email: sales@ottoenvironmental.
  com
Web: www.ottoenvironmental.
  com

research animal diagnostic laboratories

Abaxis
3240 Whipple Rd.
Union City, CA 94587 USA
Telephone: 510-675-6500
Fax: 510-441-6150
Email: abaxis@abaxis.com
Website: abaxis.com/veterinary

Ani Lytics, Inc.
200 Girard St., Ste. 200
Gaithersburg, MD 20877 USA
Toll Free: 800-237-2815
Telephone: 301-921-0168
Fax: 301-977-0433
Email: anilytics@mindspring.
  com

AnLab Ltd
142 20 Prague 4, Vídeňská 1083
Czech Republic
Telephone: + 420 261711667
Fax: +420 261711719
Email: info@anlab.cz

Antech Diagnostics GLP
507 Airport Blvd., Ste. 113
Morrisville, NC 27560 USA
Telephone: 919-787-9528
Fax: 919-277-0825
Website: www.antechglp.com

BioReliance® by SAFC
14920 Broschart Rd.
Rockville, MD 20850 USA
Toll Free: 800-553-5372
Telephone: 301-738-1000
Fax: 301-610-2590
Email: info@bioreliance.com
Website: www.bioreliance.com

BioLASCO Taiwan Co., Ltd.
3F, 316, Chung-Yang Rd.
Taipei 115 Taiwan
Telephone: 886-2-2651-8669
Fax: 886-2-2651-8969
Email: lasco@biolasco.com.tw
Website: www.biolasco.com.tw

Biotech Trading Partners
1042B N. El Camino Real, Ste. 341
Encinitas, CA 92024 USA
Telephone: 760-578-6176
Fax: 267-295-8218
Website: www.biotech-central.com

Charles River Laboratories
251 Ballardvale St.
Wilmington, MA 01887 USA
Toll Free: 800-522-7287
Telephone: 978-658-6000
Fax: 978-988-9236
Email: comments@crl.com
Website: www.criver.com

Colonial Medical Supply Co., Inc.
PO Box 866
Franconia, NH 03580-0866 USA
Toll Free: 888-446-8427
Telephone: 603-823-9911
Fax: 603-823-8799
Email: cms@colmedsupply.com
Website: www.colmedsupply.com

IDEXX Laboratories, Inc.
2825 Kovr Dr.
West Sacramento, CA 95605 USA
Telephone: 916-731-4017
Fax: 916-372-2783
Email: kristina-robertson@idexx.com
Website: www.idexx.com/preclinicalresearch

IDEXX RADIL™
Lab Animal and Biological Materials Diagnostic Testing
4011 Discovery Dr.
Columbia, MO 65201 USA
Toll Free: 800-669-0825
Telephone: 573-499-5700
Fax: 573-499-5701
Email: idexx-radil@idexx.com
Website: www.radil.missouri.edu

The Microbiology Laboratories
56 Northumberland Rd.
North Harrow, Middlesex HA2
 7RE UK
Telephone: +44 2088 684050
Fax: +44 2088 666100
Email: needham@microlabs.
 demon.co.uk

**Lampire Biological
Laboratories**
PO Box 270
Pipersville, PA 18947 USA
Telephone: 215-795-2838
Fax: 215-795-0237
Email: lampire@lampire.com
Website: www.lampire.com

research support

Access Technologies
7350 N. Ridgeway
Skokie, IL 60076 USA
Telephone: 847-674-7131
Fax: 847-674-7066
Email: info@norfolk.com
Website: www.norfolkaccess.
 com

**ALZET Osmotic Pumps/
DURECT Corp.**
2 Results Way
Cupertino, CA 95014 USA
Telephone: 408-367-4036
Fax: 408-865-1406
Email: alzet@durect.com
Website: www.alzet.com

**Animal Care Training
Services**
340 Cypress Cir.
King Of Prussia, PA 19406 USA
Telephone: 609-346-9945
Email: info@actstraining.com
Website: www.actstraining.com

**Animal Identification &
Marking Systems Inc (AIMS)**
99 Park Dr.
Hornell, NY 14843 USA
Telephone: 607-324-6752
Fax: 607-324-6753
Email: aims@animalid.com
Website: www.animalid.com

**AVID Identification
Systems, Inc.**
3185 Hamner Ave.
Norco, CA 92860 USA
Telephone: 951-371-7505
Fax: 951-737-8967
Email: sales@avidid.com
Website: www.avidid.com

BASi
2701 Kent Ave.
West Lafayette, IN 47906 USA
Telephone: 765-463-4527
Fax: 765-497-1102
Email: basi@basinc.com
Website: www.basinc.com

Best Theratronics
413 March Rd.
Ottawa, ON K2K 0E4 Canada
Telephone: 613-591-2100
Fax: 613-591-6627
Email: info@theratronics.ca
Website: www.theratronics.ca

BigC/Dino-Lite Scopes
20655 S. Western Ave., Ste. 116
Torrance, CA 90501 USA
Telephone: 310-618-9990
Fax: 310-618-9996
Email: sales@bigc.com
Website: www.bigc.com

BioDAQ
20 Jules Ln.
New Brunswick, NJ 08901 USA
Toll Free: 877-486-2486
Telephone: 732-247-2390
Fax: 732-247-2340
Email: info@researchdiets.com
Website: www.researchdiets.
   com/biodaq

Biopticon Corporation
200 Federal St., Ste. 223
Camden, NJ 08103 USA
Telephone: 609-275-0321
Fax: 856-225-6683
Email: info@biopticon.com
Website: www.biopticon.com

BioMedic Data Systems, Inc.
1 Silas Rd.
Seaford, DE 19973 USA
Telephone: 302-628-4100
Fax: 302-628-4110
Email: inquiry@bmds.com
Website: www.bmds.com

BioVision Veterinary Endoscopy
211 Corporate Cir., Suite H
Golden, CO 80401 USA
Toll Free: 877-304-3738
Telephone: 303-225-0960
Fax: 303-237-0757
Email: info@biovisiontech.com
Website: www.biovisiontech.com

Braintree Scientific
PO Box 361
Braintree, MA 02185 USA
Telephone: 781-348-0768
Fax: 781-843-7932
Email: info@braintreesci.com
Website: www.braintreesci.com

Crist Instrument Co., Inc.
111 W. First St.
Hagerstown, MD 21740 USA
Telephone: 301-393-8615
Fax: 301-393-8618
Website : www.cristinstrument.
   com

Covance Research Products, Inc.
PO Box 7200
Denver, PA 17517 USA
Telephone: 717-336-4921
Fax: 717-336-5344
Email: info@covance.com
Website: www.covance.com

Data Sciences International (DSI)
119 14th St., NW
Saint Paul, MN 55112 USA
Telephone: 651-481-7400
Fax: 651-481-7417
Email: inforrnation@datasci.com
Website: www.datasci.com

**Diagnostic Imaging
Systems, Inc.**
PO Box 3390
Rapid City, SD 57709 USA
Telephone: 605-341-2433
Fax: 605-341-0053
Email: sales@vetxray.com
Website: www.vetxray.com

DiLab CMA Microdialysis, Inc.
73 Princeton St.
North Chelmsford, MA 01863
    USA
Toll Free: 800-440-4980
Telephone: 978-251-1950
Fax: 978-251-1950
Email: dilab.usa@microdialysis.
    com
Website: www.microdialysis.se

DRE Veterinary
1800 Williamson Ct.
Louisville, KY 40223 USA
Toll Free: 800-979-6795
Telephone: 502-244-6345
Fax: 502-244-0369
Email: info@dremedical.com
Website: www.dreveterinary.com

Drew Scientific, Inc.
4230 Shilling Way
Dallas, TX 75237 USA
Telephone: 214-210-4900
Fax: 214-210-4949
Email: vetproducts@drew-
    scientific.com
Website: www.drew-scientific.
    com

Esco Micro Pte., Ltd.
21 Changi St. 1
Singapore 486777
Telephone: +65 6542-0833
Fax: +65 6542-6920
Email: mail@escoglobal.com
Website: www.escoglobal.com
Facebook: www.facebook.com/
    pages/Esco/115655397163
Twitter: twitter.com/escoglobal

E-Z Systems Euthanex Corp.
PO Box 3544
Palmer, PA 18043 USA
Telephone: 610-559-0159
Fax: 610-821-3061
Email: info@euthanex.com
Website: www.euthanex.com

Faxitron Bioptics, LLC
3440 E. Britannia Dr.,
    Ste. 150 B
Tucson, AZ 85706 USA
Telephone: 520-399-8180
Fax: 520-399-8182
Website: www.faxitron.com

First Biomedical, Inc.
878 N. Jan-Mar Ct.
Olathe, KS 66061 USA
Toll Free: 800-658-5582
Fax: 913-764-5282
Email: sales@firstbiomed.com
Website: www.firstbiomedical.
    com

Glenbrook Technologies, Inc.
11 Emery Ave.
Randolph, NJ 07869 USA
Telephone: 973-361-8866
Fax: 973-361-9286
Website: www.glenbrooktech.com

IITC Life Science, Inc.
23924 Victory Blvd.
Woodland Hills, CA 91367 USA
Telephone: 818-710-8843
Fax: 818-992-5185
Email: iitc@iitcinc.com
Website: www.iitcinc.com

Instech Solomon
5209 Militia Hill Rd.
Plymouth Meeting, PA 19462
   USA
Toll Free: 800-443-4227
Telephone: 610-941-0132
Fax: 610-941-0134
Website: www.instechlabs.com
Twitter: twitter.com/instechlabs

InterMetro Industries
651 N. Washington St.
Wilkes-Barre, PA 18705 USA
Toll Free: 800-992-1776
Telephone: 570-825-2741
Fax: 570-825-2852
Email: moreinfo-cp@metro.com
Website: www.metro.com

JL Shepherd & Associates
1010 Arroyo St.
San Fernando, CA 91340 USA
Telephone: 818-898-2361
Fax: 818-361-8095
Email: info@jlshepherd.com
Website: www.jlshepherd.com

Ketchum Manufacturing, Inc.
1245 California Ave.
Brockville, ON K6V 7N5
   Canada
Telephone: 613-342-8455
Fax: 613-342-7550
Email: ketchum@sympatico.ca
Website: www.ketchum.ca

Key Surgical, Inc.
8101 Wallace Rd.
Eden Prairie, MN 55344 USA
Telephone: 952-914-9789
Fax: 952-914-9866
Email: info@keysurgical.com
Website: www.keysurgical.com

Lab Etc, Inc.
PO Box 111
Clayton, DE 19938 USA
Toll Free: 877-452-2382
Fax: 302-338-8822
Email: fish@labetc.net
Website: www.labetc.net

Leica Microsystems, Inc.
1700 Leider Ln.
Buffalo Grove, IL 60089 USA
Telephone: 800-248-0123
Fax: 847-405-0164
Website: www.leica-
   microsystems.com

Med Associates, Inc.
PO Box 319
St. Albans, VT 05478 USA
Telephone: 802-527-9724
Fax: 802-524-2110
Email: info@med-associates.com
Website: www.med-associates.
   com

Microchip ID
PO Box 1028
Lake Zurich, IL 60047 USA
Telephone: 847-387-0577
Fax: 847-726-7835
Email: sales@microchipid.us
Website: www.microchipid.us

Molecular Imaging Products Company
63043 Lower Meadow Dr.,
    Unit 110
Bend, OR 97701 USA
Telephone: 541-385-9967
Fax: 541-385-9662
Email: info@mipcompany.com
Website: www.mipcompany.com

Plas-Labs, Inc.
401 E. North St.
Lansing, MI 48906 USA
Toll Free: 800-866-7527
Telephone: 517-372-7177
Fax: 517-372-2857
Email: sales@plas-labs.com
Website: www.plas-labs.com

RadSource Technologies, Inc.
480 Brogdon Rd., Ste. 500
Suwanee, GA 30024 USA
Telephone: 770-887-8669
Fax: 678-302-7510
Email: info@radsource.com
Website: www.radsource.com

Roboz Surgical Instrument Co., Inc.
PO Box 10710
Gaithersburg, MD 20898 USA
Toll Free: 800-424-2984
Fax: 888-424-3121
Email: info@roboz.com
Website: www.roboz.com

S & B Medvet GmbH
Neuer Weg 4 D
Babenhausen, 64832 Germany
Telephone: +49 6073-725836
Fax: +49 6073-725831
Email: medvet@t-online.de
Website: www.submedvet.com

Sable Systems International, Inc.
6000 S. Eastern Ave., Bldg. 1
Las Vegas, NV 89119 USA
Telephone: 702-269-4445
Fax: 702-269-4446
Email: mail@sablesys.com
Website: www.sablesys.com

SAI Strategic Applications
1655 Northwind Blvd.
Libertyville, IL 60048 USA
Telephone: 847-680-9385
Fax: 847-680-9837
Email: info@sai-infusion.com
Website: www.sai-infusion.com

Sarstedt
PO Box 468
Newton, NC 28658 USA
Telephone: 828-465-4000
Fax: 828-465-0718
Website: www.sarstedt.com/
    php/main.php

Transnetyx, Inc.
8110 Cordova Rd., Ste. 119
Cordova, TN 38016 USA
Toll Free: 888-321-2113
Telephone: 901-507-0476
Fax: 901-507-0480
Email: customerservice@
    transnetyx.com
Website: www.transnetyx.com

VetEquip, Inc.
PO Box 10785
Pleasanton, CA 94588 USA
Toll Free: 800-466-6463
Telephone: 925-463-1828
Fax: 925-463-1943
Email: info@vetequip.com
Website: www.vetequip.com

sanitation supplies

The Baker Company
PO Box E
Sanford, ME 04073 USA
Telephone: 207-324-8773
Fax: 207-324-3869
Website: www.bakerco.com

Betterbuilt
1388 Derwent Way
Delta, BC V3M 6C4 Canada
Telephone: 604-777-9988
Fax: 604-777-9910
Email: info@nsc-betterbuilt.com
Website: www.nsc-betterbuilt.com

BioLASCO Taiwan Co., Ltd.
3F, 316, Chung-Yang Rd.
Taipei 115 Taiwan
Telephone: 886-2-2651-8669
Fax: 886-2-2651-8969
Email: lasco@biolasco.com.tw
Website: www.biolasco.com.tw

Bioquell, Inc.
101 Witmer Rd., Ste. 500
Horsham, PA 19044 USA
Telephone: 215-682-0225
Fax: 215-682-0395
Website: www.bioquell.com

BioSAFE Engineering WR2
485 Southpoint Cir.
Brownsburg, IN 46112 USA
Telephone: 317-858-8099
Fax: 317-858-8202
Email: info@biosafeengineering.
    com
Website : www.
    biosafeengineering.com

CANI, Inc.
1758 Allentown Rd., PMB 146
Lansdale, PA 19446 USA
Telephone: 610-222-4500
Fax: 610-222-4203
Email: caniusa@caniusa.com
Website: www.caniusa.com

ClorDiSys Solutions, Inc.
PO Box 549
Lebanon, NJ 08833 USA
Telephone: 908-236-4100
Fax: 908-236-2222
Email: info@clordisys.com
Website: www.clordisys.com

Consolidated Sterilizer Systems
76 Ashford St.
Boston, MA 02134 USA
Telephone: 617-782-6072
Fax: 617-787-5865
Email: info@consteril.com
Website: www.consteril.com

Contec, Inc.
525 Locust Grove
Spartanburg, SC 29303 USA
Telephone: 864-503-8333
Fax: 864-503-8444
Email: wipers@contecinc.com
Website: www.contecinc.com

DetachAB
Sundbyvägen 24
Strängsäs, 64551 Sweden
Telephone: +46 152-228 30
Fax: +46 152-130 80
Email: info@detach.com
Website: www.detach.com

DuPont Animal Health Solutions
1007 Market St.
Wilmington, DE 19898 USA
Toll Free: 888-243-4608
Telephone: 302-892-7575
Fax: 302-892-7656
Email: biosecurity@gbr.dupont.
    com
Website: www.ahs.dupont.com

Dustcontrol, Inc.
6720 Amsterdam Way
Wilmington, NC 28405 USA
Telephone: 910-395-1808
Fax: 910-395-2110
Email: sales@dustcontrolusa.
    com
Website: www.transmaticgroup.
    com

Esco Micro Pte, Ltd.
21 Changi St. 1
Singapore 486777
Telephone: +65 6542-0833
Fax: +65 6542-6920
Email: mail@escoglobal.com
Website: www.escoglobal.com
Facebook: www.facebook.com/
    pages/Esco/115655397163
Twitter: twitter.com/escoglobal

ETC
125 James Way
Southampton, PA 18966 USA
Telephone: 215-355-9100
Fax: 215-357-4000
Email: sterilizers@etcusa.com
Website: www.etcsterilization.
    com

Garner Decontamination Solutions
1717 W. 13th St.
Deer Park, TX 77536 USA
Telephone: 281-930-1200
Fax: 281-478-0296
Website: www.
    garnerdecontamination
    solutions.com

Germfree
11 Aviator Way
Ormond Beach, FL 32174 USA
Telephone: 386-677-7742
Fax: 386-677-0442
Email: info@germfree.com
Website: www.germfree.com

Getinge
1777 E. Henrietta Rd.
Rochester, NY 14623 USA
Telephone: 585-475-1400
Fax: 585-272-5116
Email: info@getingeusa.com
Website: www.getinge.com

Girton Manufacturing Co, Inc.
PO Box 900
Millville, PA 17846 USA
Telephone: 570-458-5521
Fax: 570-458-5589
Email: info@girton.com
Website: www.girton.com

Gruenberg Steri-Dry
2821 Old Route 15
New Columbia, PA 17856 USA
Telephone: 570-538-7227
Fax: 570-538-7380
Email: tpsinfo@tsp.spx.com
Website: www.gruenberg.com

Hapman
6002 E. Kilgore Rd.
Kalamazoo, MI 49008 USA
Telephone: 780-430-4225 x223
Fax: 269-349-2477
Website: www.hapman.com

International Enviroguard
2400 Skyline Dr., Ste. 400
Mesquite, TX 75149 USA
Telephone: 214-388-4012
Fax: 214-388-5839
Website: www.int-enviroguard.
com

Kimberly-Clark Professional
1400 Holcomb Bridge Rd.
Roswell, GA 30076 USA
Telephone: 770-587-8000
Fax: 770-587-7676
Email: info@kimtech.com
Website: www.kimtech.com

Lynx Products Group, LLC
650 Lake St.
Wilson, NY 14172 USA
Telephone: 716-751-3100
Fax: 716-751-3101
Email: info@lynxpg.com
Website: www.lynxpg.com

Mar Cor Purification
14550 28th Ave. North
Plymouth, MN 55447 USA
Toll Free: 800-633-3080
Telephone: 484-991-0220
Fax: 763-210-3868
Email: info@mcpur.com
Website: www.mcpur.com

Micronova Manufacturing, Inc.
3431 W. Lomita Blvd.
Torrance, CA 90505 USA
Telephone: 310-784-6990
Fax: 310-784-6980
Email: info@micronova-mfg.com
Website: www.micronova-mfg.
com

NuAire, Inc.
2100 Fernbrook Ln.
Plymouth, MN 55447 USA
Telephone: 763-553-1270
Fax: 763-553-0459
Email: nuaire@nuaire.com
Website: www.nuaire.com

Pharmacal Research Labs
PO Box 369
Naugatuck, CT 06770 USA
Toll Free: 800-243-5350
Telephone: 203-755-4908
Fax: 203-755-4309
Email: moreinfo@pharmacal.
   com
Website: www.pharmacal.com

**PRIMUS Sterilizer
Company, LLC**
117 South 25th St.
Omaha, NE 68131 USA
Telephone: 402-344-4200
Fax: 402-344-4242
Website: www.primus-sterilizer.
   com

Progressive Recovery, Inc.
700 Industrial Dr.
Dupo, IL 62239 USA
Telephone: 618-286-5000
Fax: 618-286-5009
Website: www.pri-bio.com

Quip Laboratories, Inc.
1500 Eastlawn Ave.
Wilmington, DE 19802 USA
Toll Free: 800-424-2436
Telephone: 302-761-2600
Fax: 302-761-2611
Email: quip@quiplabs.com
Website: www.quiplabs.com

**R-V Industries, Inc.
Beta Star Equipment**
584 Poplar Rd.
Honey Brook, PA 19344 USA
Telephone: 610-273-2457
Fax: 610-273-3361
Email: sales@rvii.com
Website: www.rvii.com

Resurgent Health and Medical
600 Corporate Cir., Ste. H
Golden, CO 80401 USA
Toll Free: 800-932-7707
Fax: 303-790-4859
Email: cleantech@meritech.com
Website: www.resurgenthealth.
   com

Rochester Midland Corp.
333 Hollenbeck St.
Rochester, NY 14621 USA
Toll Free: 800-836-1627
Telephone: 585-336-2376
Fax: 716-336-2357
Website: www.rochestermidland.
   com

Schlyer Machine
814 Wurlitzer Dr.
North Tonawanda, NY 14120 USA
Telephone: 716-696-3171
Fax: 716-696-3174
Website: www.schlyermachine.
   com

**Scientek Technology
Corporation**
7943 Progress Way
Delta, BC V4G 1A3 Canada
Telephone: 604-940-8084
Fax: 604-940-8085
Email: sales@scientek.net
Website: www.scientek.net

SPX® Thermal Product Solutions
PO Box 150
White Deer, PA 17887 USA
Telephone: 570-538-7200
Toll frcc: 800-586-2473
Email: tps.info@spx.com
Website: www.
    thermalproductsolutions.com

STERIS Corp.
5960 Heisley Rd.
Mentor, OH 44060 USA
Toll Free: 800-444-9009
Telephone: 440-354-2600
Fax: 440-357-2321
Website: www.steris.com

Sychem, Ltd.
Highcove House, Victory Close,
    Chandler's Ford
Eastleigh, Hampshire SO53 4BD
    UK
Telephone: +44 845 644 6824
Fax: +44 2380 269016
Email: sales@sychem.co.uk
Website: www.sychem.co.uk

Tuttnauer Co., Ltd.
25 Power Dr.
Hauppauge, NY 11788 USA
Toll Free: 800-624-5836
Fax: 631-737-1034
Email: info@tuttnauer.com
Website: www.tuttnauer.com

transportation

Airnet Express
7215 Star Check Dr.
Columbus, OH 43217 USA
Toll Free: 888-888-8463
Fax: 614-409-7852
Email: solutions@airnet.com
Website: www.airnet.com

Frames Animal Transport, Inc.
1119 Haverford
Ridley Park, PA 19078 USA
Telephone: 610-399-5166
Fax: 610-399-5165

Marken
521 RXR Plz.
East Tower
Uniondale, NY 11556 USA
Toll Free: 800-932-6755
Telephone: 516-307-3287
Website: www.marken.com

MNX-Midnite Express Global Logistics
300 N. Oak St.
Los Angeles, CA 90302 USA
Toll Free: 800-643-6483
Telephone: 310-330-2300
Fax: 310-330-2358
Website: www.mnx.com

**Transport Container
Corporation**
PO Box 163183
Columbus, OH 43216 USA
Telephone: 614-459-8140
Fax: 614-459-9165

World Courier
Toll Free: 888-221-6600
Website: www.worldcourier.com

veterinary and surgical supplies

Abbeyvet Export
Sherburn Enterprise Park
Aviation Way
Sherburn-in-Elmet
Nr. Leeds
North Yorkshire, England
LS25 6NB
Telephone: +44 1977 685777
Fax: +44 1977 685111
Email: enquiries@abbeyvet-
    export.co.uk
Website: www.abbeyvet-export.
    co.uk

Abbott Animal Health
200 Abbott Park Rd.
Abbott Park, IL 60064 USA
Toll Free: 888-299-7416
Fax: 847-938-2741
Website: www.
    abbottanimalhealth.com

Andersen Products, Inc.
3202 Caroline Dr.
Haw River, NC 27258 USA
Telephone: 336-376-3000
Fax: 336-376-8153
Email: mailbox@anpro.com
Website: www.anpro.com

Bio-Serv
One 8th St.
Frenchtown, NJ 08825 USA
Telephone: 908-996-2155
Fax: 908-996-4123
Toll Free: 800-996-9908
Email: sales@bio-serv.com
Website: www.bio-serv.com

**BioVision Veterinary
Endoscopy**
211 Corporate Cir., Ste. H
Golden, CO 80401 USA
Toll Free: 877-304-3738
Telephone: 303-225-0960
Fax: 303-237-0757
Email: info@biovisiontech.com
Website: www.biovisiontech.com

Braintree Scientific
PO Box 361
Braintree, MA 02185 USA
Telephone: 781-348-0768
Fax: 781-843-7932
Email: info@braintreesci.com
Website: www.braintreesci.com

Butler Schein Animal Health
400 Metro Pl. North
Dublin, OH 43017 USA
Telephone: 614-659-1754
Fax: 888-329-3861
Website: www.butlerschein.com

Colonial Medical Supply Co., Inc.
PO Box 866
Franconia, NH 03580-0866 USA
Telephone: 603-823-9911
Fax: 603-823-8799
Toll Free: 888-446-8427
Email: cms@colmedsupply.com
Website: www.colmedsupply.com

Crist Instrument Co., Inc.
111 W. First St.
Hagerstown, MD 21740 USA
Telephone: 301-393-8615
Fax: 301-393-8618
Website: www.cristinstrument.
   com

Diagnostic Imaging Systems, Inc.
PO Box 3390
Rapid City, SD 57709 USA
Telephone: 605-341-2433
Fax: 605-341-0053
Email: sales@vetxray.com
Website: www.vetxray.com

Glenbrook Technologies, Inc.
11 Emery Ave.
Randolph, NJ 07869 USA
Telephone: 973-361-8866
Fax: 973-361-9286
Website: www.glenbrooktech.
   com

Jorgensen Laboratories
1450 N. Van Buren Ave.
Loveland, CO 80538 USA
Telephone: 970-669-2500
Fax: 970-663-5042
Email: info@jorvet.com
Website: www.jorvet.com

Ketchum Manufacturing, Inc.
1245 California Ave.
Brockville, ON K6V 7N5 Canada
Telephone: 613-342-8455
Fax: 613-342-7550
Email: ketchum@sympatico.ca
Website: www.ketchum.ca

Key Surgical, Inc.
8101 Wallace Rd.
Eden Prairie, MN 55344 USA
Telephone: 952-914-9789
Fax: 952-914-9866
Email: info@keysurgical.com
Website: www.keysurgical.com

Lomir Biomedical, Inc.
95 Huot
Notre Dame De l'Ile
Perrot, QC J7V 7M4 Canada
Telephone: 514-425-3604
Fax: 514-425-3605
Email: info@lomir.com
Website: www.lomir.com

MEDIpoint, Inc.
72 E. 2nd St.
Mineola, NY 11501 USA
Telephone: 516-294-8822
Fax: 516-746-6693
Website: www.medipoint.com

Meds for Vets
585 W. 9400 S., Ste. 100
Sandy, UT 84070 USA
Telephone: 801-255-7666
Fax: 801-255-7690
Email: info@medsforvets.com
Website: www.medsforvets.com

Midwest Veterinary Supply
11965 Larc Industrial Blvd.
Burnsville, MN 55337 USA
Telephone: 952-894-4350
Fax: 952-894-5407
Website: www.midwestvet.net

MiteArrest
56 Hawes St.
Brookline, MA 02446 USA
Telephone: 617-742-2400
Fax: 617-849-5494
Email: info@mitearrest.com
Website: www.mitearrest.com

Molecular Imaging Products
63043 Lower Meadow Dr.,
   Unit 110
Bend, OR 97701 USA
Telephone: 541-385-9967
Fax: 541-385-9662
Email: info@mipcompany.com
Website: www.mipcompany.com

Mopec, Inc.
21750 Coolidge Hwy.
Oak Park, MI 48237 USA
Telephone: 248-284-0890
Fax: 248-291-2050
Email: info@mopec.com
Website: www.mopec.com

Roboz Surgical Instrument Co., Inc.
PO Box 10710
Gaithersburg, MD 20898 USA
Toll Free: 800-424-2984
Fax: 888-424-3121
Email: info@roboz.com
Website: www.roboz.com

S & B Medvet GmbH
Neuer Weg 4 D
Babenhausen, 64832 Germany
Telephone: +49 6073-725836
Fax: +49 6073-725831
Email: medvet@t-online.de
Website: www.submedvet.com

Tuttnauer Co., Ltd.
25 Power Dr.
Hauppauge, NY 11788 USA
Toll Free: 800-624-5836
Fax: 631-737-1034
Email: info@tuttnauer.com
Website: www.tuttnauer.com

VetEquip, Inc.
PO Box 10785
Pleasanton, CA 94588 USA
Toll Free: 800-466-6463
Telephone: 925-463-1828
Fax: 925-463-1943
Email: info@vetequip.com
Website: www.vetequip.com

VetaPharma, Ltd.
Sherburn-in-Elmet
Leeds, UK
Telephone: +44 (0)1977-680100
Fax: +44 (0)1977-685111
Website: www.vetapharma.co.uk

bibliography

chapter 1

Addison, WHF, Appleton, JL. 1915. Structure and growth of rat incisors. *J Morph* 26: 43–96.

Argent, NB. et al. 1994. A new method for measuring the blood volume of the rat using [113m]indium as a tracer. *Lab Anim* 28(2): 172–175.

Asano, Y. et al. 1998. Haematological and serum biochemical values in spontaneously epileptic male rats and related rat strains. *Lab Anim* 32(2): 214–218.

Augustsson, H. et al. 2002. Human–animal interactions and animal welfare in conventionally and pen-housed rats. *Lab Anim* 36(3): 271–281.

Avsaroglu, H. et al. 2007. Differences in response to anaesthetics and analgesics between inbred rat strains. *Lab Anim* 41(3): 337–344.

Azar, TA. et al. 2005. Stress-like cardiovascular responses to common procedures in male versus female spontaneously hypertensive rats. *Contemp Top Lab Anim Sci* 44(3): 25–30.

Baker-Herman, TL. et al. 2010. Differential expression of respiratory long-term facilitation among inbred rat strains. *Respir Physiol Neurobiol* 170(3): 260–267.

Balcombe, JP. 2006. Laboratory environments and rodents' behavioural needs: A review. *Lab Anim* 40: 217–235.

Bass, NH, Lundborg, P. 1973. Postnatal development of bulk flow in the cerebrospinal fluid system of the albino rat: Clearance of carboxyl-[14c]inulin after intrathecal infusion. *Brain Res* 52: 323–332.

Benus, RF. et al. 1991. Heritable variation for aggression as a reflection of individual coping strategies. *Experientia* 47(10): 1008–1019.

Bjork, E. et al. 2000. R-weighting provides better estimation for rat hearing sensitivity. *Lab Anim* 34(2): 36–144.

Blanchard, RJ. et al. 2001. Animal models of social stress: Effects on behavior and brain neurochemical systems. *Physiol Behav* 73(3): 261–271.

Borg, TK. et al. 1982. Functional arrangement of connective tissue in striated muscle with emphasis on cardiac muscle. *Scan Electron Microsc* (Pt 4): 1775–1784.

Brewer, NR, Cruise, LJ. 2000. Mammalian thermoregulation: Species differences. *Contemp Top Lab Anim Sci* 39(3): 23–27.

Burda, H, Voldrick, I. 1980. Correlation between the hair cell density and the auditory threshold in the white rat. *Hear Res* 3(1): 91–93.

Bures, J. et al. 1976. *Techniques and Basic Experiments for the Study of Brain and Behavior*. Amsterdam: Elsevier.

Burn, CC. 2008. What is it like to be a rat? Rat sensory perception and its implications for experimental design and rat welfare. *Appl Anim Behav Sci* 112(1): 1–32.

Caligioni, CS. 2009. Assessing reproductive status/stages in mice. *Curr Protoc Neurosci* Appendix 4I.

Callicott, RJ. et al. 2007. Genomic comparison of Lewis and Wistar-Furth Rat substrains by use of microsatellite markers. *J Am Assoc Lab Anim Sci* 46(2): 25–29.

Canzian, F. 1997. Phylogenetics of the laboratory rat *Rattus norvegicus*. *Genome Res* 7: 262–267.

Car, BD. et al. 2006. Clinical pathology of the rat. In: Suckow MA, Weisbroth SH, Franklin CL, editors. *The Laboratory Rat*, 2nd ed. Elsevier, pp. 127–146.

Carvell, GE, Simons, DJ. 1996. Abnormal tactile experience early in life disrupts active touch. *J Neurosci* 16(8): 2750–2757.

Cavigelli, SA. et al. 2006. Fecal corticoid metabolites in aged male and female rats after husbandry-related disturbances in the colony room. *J Am Assoc Lab Anim Sci* 45(6): 17–21.

Cesar, S. et al. 2003. Left ventricular end-diastolic pressure-volume relationship in septic rats with open thorax. *Comp Med* 53(5): 493–497.

Chew, RM. 1968. Water metabolism in mammals. *Physiol Mammal* 2: 143–178.

Christiansen, T. et al. 1997. Relationship between MRI and morpho-metric kidney measurements in diabetic and non-diabetic rats. *Kidney Int* 51(1): 50–56.

Davies, B, Morris, T. 1993. Physiological parameters in laboratory animals and humans. *Pharm Res* 10(7): 1093–1095.

Davis, LDR. et al. 2011. Maternal separation and gastrointestinal transit time in neonate rats. *Lab Anim* 45(4): 280–282.

DePasquale, M. et al. 1989. Brain ion and volume regulation during acute hypernatremia in Battleboro rats. *Am J Physiol* 256 (6 Pt 2), F1059–FI066.

DeSesso, JM, Jacobson, CF. 2001. Anatomical and physiological parameters affecting gastrointestinal absorption in humans and rats. *Food Chem Toxicol* 39(3): 209–228.

Derian, CK. et al. 1995. Species Differences in platelet responses to thrombin and SFLLRN: Receptor-mediated calcium mobilization and aggregation, and regulation by protein kinases. *Thromb Res* 78(6): 505–519.

Durand, G. et al. 2001. Spontaneous polar anterior subcapsular len-ticular opacity in Sprague-Dawley rats. *Comp Med* 51(2): 176–179.

D'Uscio, LV. et al. 2000. Circulation. In: Krinke GJ, editor. *The Laboratory Rat*, 2nd ed. London: Academic Press, pp. 345–357.

Erben, RG. et al. 2000. Androgen deficiency induces high turnover osteopenia in aged male rats: A sequential histomorphometric study. *J Bone Miner Res* 15(6): 1085–1098.

Fomby, LM. et al. 2004. Use of CO_2/O_2 anesthesia in the collection of samples for serum corticosterone analysis from Fischer 344 rats. *Contemp Top Lab Anim Sci* 43(2): 8–12.

França, LR. et al. 1998. Germ cell genotype controls cell cycle during spermatogenesis in the rat. *Biol Reprod* 59(6): 1371–1377.

Gill, TJ, III. et al. 1992. Definition, nomenclature, and conservation of the rat strains. *ILAR News* 34: S1–S26.

Gill, TJ. et al. 1989. The rat as an experimental animal. *Science* 245(4915): 269–276.

Goicoechea, M. et al. 2008. Minimizing creatine kinase variability in rats for neuromuscular research purposes. *Lab Anim* 42(1): 19–25.

Goldblum, D. et al. 2002. Non-invasive determination of intraocular pressure in the rat eye: Comparison of an electronic tonometer (TonoPen®), and a rebound (impact probe) tonometer. *Graefes Arch Clin Exp Ophthalmol* 240(11): 942–946.

Gould, EM. 2008. The effect of ketamine/xylazine and carbon dioxide on plasma luteinizing hormone releasing hormone and testosterone concentrations in the male Norway rat. *Lab Anim* 42(4): 483–488.

Goyal, HO. et al. 2005. Permanent induction of morphological abnormalities in the penis and penile skeletal muscles in adult rats treated neonatally with diethylstilbestrol or estradiol valerate: A dose-response study. *J Androl* 26(1): 32–43.

Gur, E, Waner, T. 1993. The variability of organ weight background data in rats. *Lab Anim* 27(1): 65–72.

Hafez, E. 1970. *Reproduction and Breeding Techniques for Laboratory Animals*. Philadelphia: Lea & Febiger.

Hall, HD, Schneyer, CA. 1964. Paper electrophoresis of rat salivary secretions. *Proc Soc Exp Biol Med* 115: 1001–1005.

Hammer, RE. et al. 1990. Spontaneous inflammatory disease in transgenic rats expressing hla-b27 and human β2m: An animal model of hla-b27-associated human disorders. *Cell* 63(5): 1099–1112.

Harkness E, Wagner JE. 1995. *The Biology and Medicine of Rabbits and Rodents*. Baltimore: Williams & Wilkins.

Hatakeyama, S. et al. 1987. A sexual dimorphism of mucous cells in the submandibular salivary gland of Rat. *Arch Oral Biol* 32(10): 689–693.

Heffner, HE, Heffner, RS. 2007. Hearing ranges of laboratory animals. *J Am Assoc Lab Anim Sci* 46(1): 20–22.

Heffner, HE. et al. 1994. Audiogram of the Hooded Norway rat. *Hear Res* 73(2): 244–247.

Heffner, RS, Heffner, HE. 1992. Visual factors in sound localization in mammals. *J Comp Neurol* 317(3): 219–232.

Heiderstadt, KM. et al. 2000. The effect of chronic food and water restriction on open-field behaviour and serum corticosterone levels in rats. *Lab Anim* 34(1): 20–28.

Hofstetter, J. et al. 2006. Morphophysiology. In: Suckow MA, Weisbroth SH, Franklin CL, editors. *The Laboratory Rat*, 2nd ed. Elsevier, pp. 94–125.

Hojo, H. et al. 2001. Abnormalities of gonadotrophs in the adenohypophysis of infertile male PD rats. *Comp Med* 51(5): 462–466.

Ito, K. et al. 2001. The effect of food consistency and dehydration on reflex parotid and submandibular salivary secretion in conscious rats. *Arch Oral Biol* 46(4): 353–363.

Jacobs, GH. et al. 2001. Cone-based vision of rats for ultraviolet and visible lights. *J Exp Biol* 204(Pt 14): 2439–2446.

Kaneko, JJ. et al. 2008. *Clinical Biochemistry of Domestic Animals*, 6th ed. Burlington (MA): Academic Press.

Kararli, TT. 1995. Comparison of the gastrointestinal anatomy, physiology, and biochemistry of humans and commonly used laboratory animals. *Biopharm Drug Dispos* 16(5): 351–380.

Kasiske, BL, Keane, WF. 2000. Laboratory assessment of renal disease: Clearance, urinalysis, and renal biopsy. In: Brenner BM, editor. *Brenner and Rector's The Kidney*. Philadelphia: W.B. Saunders.

Kilborn, SH. et al. 2002. Review of growth plate closure compared with age at sexual maturity and lifespan in laboratory animals. *Contemp Top Lab Anim Sci* 41(5): 21–26.

Kirkeby, OJ. 1991. Bone metabolism and repair are normal in athymic rats. *Acta Orthop Scand* 62(3): 253–256.

Klemetti, E. et al. 2005. Short duration hyperbaric oxygen treatment effects blood flow in rats: Pilot observations. *Lab Anim* 39(1): 116–121.

Kobayashi, M. et al. 2001. Granulomatous and cytokine responses to pulmonary cryptococcus neoformans in two strains of rats. *Mycopathologia* 151(3): 121–130.

Kohn, DF, Clifford, CB. 2002. *Biology and diseases of rats*. In: Fox JG, Anderson LC, Lowe FM, Quimby FW, editors. *Laboratory Animal Medicine*, 2nd ed. San Diego (CA): Academic Press, pp. 121–165.

Kontiola, AI. et al. 2001. The induction/impact tonometer: A new instrument to measure intraocular pressure in the rat. *Exp Eye Res* 73(6): 781–785.

Kostowsi, W. et al. 1977. Morphine action in grouped and isolated rats and mice. *Psychopharmacology (Berl)* 53(2): 191–193.

Kuiper, B. et al. 1997. Ophthalmologic examination in systemic toxicity studies: An overview. *Lab Anim* 31: 177–183.

LaBorde, JB. et al. 1999. Haematology and serum chemistry parameters of the pregnant rat. *Lab Anim* 33(3): 275–287.

Lee, KM. et al. 1998. Effects of water dilution, housing, and food on rat urine collected from the metabolism cage. *Lab Anim Sci* 48(5): 520–525.

LeFevre, J, McClintock, MK. 1988. Reproductive senescence in female rats: A longitudinal study of individual differences in estrous cycles and behavior. *Biol Reprod* 38(4): 780–789.

Leigh Perkins, LE. 2010. Preclinical models of restenosis and their application in the evaluation of drug-eluting stent systems. *Vet Pathol* 47(1): 58–76.

Leith, DE. 1976. Comparative mammalian respiratory mechanics. *Physiologist* 19(4): 485–510.

Lepschy, M. et al. 2007. Non-invasive measurement of adrenocortical activity in male and female rats. *Lab Anim* 41(3): 372–387.

Libermann, IM. et al. 1973. Blood acid–base status in normal albino rats. *Lab Anim Sci* 23(6): 862–865.

Loeb, WF, Quimby, F. 1999. *The Clinical Chemistry of Laboratory Animals*, 2nd ed. London (UK): Taylor and Francis.

Löscher, W. et al. 1998. Differences in kindling development in seven outbred and inbred rat strains. *Exp Neurol* 154: 551–559.

Lu, R. et al. 2002. Effect of age on bone mineral density and the serum concentration of endogenous nitric oxide synthase inhibitors in rats. *Comp Med* 52(3): 224–228.

Mahl, A. et al. 2000. Comparison of clinical pathology parameters with two different blood sampling techniques in rats: Retrobulbar plexus versus sublingual vein. *Lab Anim* 34(4): 351–361.

Matsuura, T. et al. 1999. Proliferative retinal changes in diabetic rats (WBN/Kob). *Lab Anim Sci* 49(5): 565–569.

McCall, RB. et al. 1969. Caretaker effect in rats. *Dev Psychol* 1(6) Part 1: 771.

Mistlberger, RE. et al. 1998. Circadian rhythms in the Zucker obese rat: Assessment and intervention. *Appetite* 30(3): 255–267.

Moore, CG. et al. 1993. Noninvasive measurement of rat intraocular pressure with the Tono-Pen. *Invest Ophthalmol Vis Sci* 34(2): 363–369.

Nahas, K. et al. 2000. Effects of acute blood removal via the sublingual vein on haematological and clinical parameters in Sprague-Dawley rats. *Lab Anim* 34(4): 362–371.

Nakamura, S, Hochwald, GM. 1983. Spinal fluid formation and glucose influx in normal and experimental hydrocephalic rats. *Experimental Neurology* 82: 108–117.

Nakayama, M. et al. 1994. Scanning electron microscopic evaluation of age-related changes in the rat vestibular system. *Otolaryngol Head Neck Surg* 111(6): 799–806.

Nuesslein-Hildesheim, B. et al. 1995. Pronounced juvenile circadian core temperature rhythms exist in several strains of rats but not in rabbits. *J Comp Physiol B* 165(1): 13–17.

O'Brien, PJ. et al. 2002. Advantages of glutamate dehydrogenase as a blood biomarker of acute hepatic injury in rats. *Lab Anim* 36(3): 313–321.

O'Brien, PJ. et al. 2006. Cardiac troponin I is a sensitive, specific biomarker of cardiac injury in laboratory animals. *Lab Anim* 40(2): 153–171.

O'Donnoghue, JM. et al. 1998. Effects of varying storage time and temperature on stability of complete blood count measurements. *Contemp Top Lab Anim Sci* 7(6): 52–54.

Ojeda, SR, Urbancki, HF. 1994. *Puberty in the rat.* In: Knobil, E, Neill JD, editors. *Physiology of Reproduction.* New York: Raven Press, pp. 363–409.

Pacha, J. 2000. Development of intestinal transport function in mammals. *Physiol Rev* 80(4): 1633–1667.

Pardridge, WM. 1991. *Peptide Drug Delivery to the Brain.* New York: Raven Press, p. 112.

Paxinos, G. 2004. *The Rat Nervous System,* 3rd ed. London: Elsevier Academic Press.

Pazirandeh, A. et al. 2004. Glucocorticoids delay age-associated thymic involution through directly affecting the thymocytes. *Endocrinology* 145(5): 2392–2401.

Percy, DH, Barthold, SW. 2007. *Pathology of Laboratory Rodents and Rabbits,* 3rd ed. Ames (IA): Blackwell, pp. 125–126.

Pérez, C. et al. 1997. Individual Housing Influences certain biochemical parameters in the rat. *Lab Anim* 31(4): 357–361.

Plećas-Solarović, B. et al. 2004. Age-dependent morphometrical changes in the thymus of male propranolol-treated rats. *Ann Anat* 186(2): 141–147.

Plećas-Solarović, B. et al. 2006. Morphometrical characteristics of age-associated changes in the thymus of old male Wistar rats. *Anat Histol Embryol* 35(6): 380–386.

Poole, TB. 1987. *The UFAW Handbook on the Care and Management of Laboratory Animals,* 6th ed. New York (NY): Churchill Livingstone.

Portfors, CV. 2007. Types and functions of ultrasonic vocalizations in laboratory rats and mice. *J Am Assoc Lab Anim Sci* 46(1): 28–34.

Portilla-De-Buen, E. et al. 2004. Activated clotting time and heparin administration in Sprague-Dawley rats and Syrian golden hamsters. *Contemp Top Lab Anim Sci* 43(2): 21–24.

Probst, RJ. et al. 2006. Gender differences in the blood volume of conscious Sprague-Dawley rat. *J Am Assoc Lab Anim Sci* 45(2): 49–52.

Rat Genome Database. [Internet]. 2011. Guidelines for nomenclature of mouse and rat strains. [Cited 23 Oct 2011]. Available at: http://rgd.mcw.edu/

Ramos, SD. et al. 2001. An inexpensive meter to measure differences in electrical resistance in the rat vagina during the ovarian cycle. *J Appl Physiol* 9(2): 667–670.

Reed, DR. et al. 2011. Body fat distribution and organ weights of 14 common strains and a 22-strain consomic panel of rats. *Physiol Behav* 103: 523–529.

Rex, A. et al. 2007. Choosing the right wild type: Behavioral and neurochemical differences between 2 populations of Sprague-Dawley rats from the same source but maintained at different sites. *J Am Assoc Lab Anim Sci* 46(5): 13–20.

Rickard, RF. et al. 2009. Characterization of a rodent model for the study of arterial microanastomoses with size discrepancy (small-to-large). *Lab Anim* 43(4): 350–356.

Robinson, R. 1965. *Genetics of the Norway Rat.* Oxford: Pergamon Press.

Rodriguez, I. et al. 1997. Mouse vaginal opening in an apoptosis-dependent process which can be prevented by the overexpression of Bcl2. *Dev Biol* 184(1): 115–121.

Rosenzweig MR. et al. 1955. Evidence for echolocation in the rat. *Science* 121(3147): 60.

Sanderson, AE. et al. 2010. Effect of cage-change frequency on rodent breeding performance. *Lab Animal* 39(6): 177–182.

Sauer, MB. et al. 2006. Clinical pathology laboratory values of rats housed in wire-bottom cages compared with those of rats housed in solid-bottom cages. *J Am Assoc Lab Anim Sci* 45(1): 30–35.

Sawaki, M. et al. 2000. Genital tract development in peripubertal female CD® IGS rats. *Comp Med* 50(3): 284–287.

Scheving, LE, Pauly JE. 1967. Daily rhythmic variations in blood coagulation times in rats. *Anat Record* 157(4): 657–666.

Schrock, H, Kuschinsky, W. 1989. Cerebrospinal fluid ionic regulation, cerebral blood flow, and glucose use during chronic metabolic acidosis. *Am J Physiol* 257(4 Pt 2), H1220–H1227.

Scott, RC. et al. 1991. The influence of skin structure on permeability: An intersite and interspecies comparison with hydrophilic penetrants. *J Invest Dermatol* 96: 921–925.

Seibel J. et al. 2010. Comparison of haematology, coagulation and clinical chemistry parameters in blood samples from the sublingual vein and vena cava in Sprague-Dawley rats. *Lab Anim* 44(4): 344–351.

Senoo, H. 2000. Digestion, metabolism. In: Krinke GJ, editor, *The Laboratory Rat.* London (UK): Academic Press, pp. 359–383.

Sharp, JL. et al. 2002. Does witnessing experimental procedures produce stress in male rats? *Contemp Top Lab Anim Sci* 41(5): 8–12.

Sharp, JL. et al. 2002. Stress-like responses to common procedures in male rats housed alone or with other rats. *Contemp Top Lab Anim Sci* 41(4): 8–14.

Sharp, JL. et al. 2002. Stress-like responses to common procedures in rats: Effect of the estrous cycle. *Contemp Top Lab Anim Sci* 41(4): 15–22.

Sharp, JL. et al. 2003. Are "by-stander" female Sprague-Dawley rats affected by experimental procedures? *Contemp Top Lab Anim Sci* 42(1): 19–27.

Sharp, JL. et al. 2003. Does cage size affect heart rate and blood pressure of male rats at rest or after procedures that induce stress-like responses? *Contemp Top Lab Anim Sci* 42(3): 8–12.

Sharp, JL. et al. 2003. Stress-like responses to common procedures in individually and group-housed female rats. *Contemp Top Lab Anim Sci* 42(1): 9–18.

Sharp, JL. et al. 2005. Selective adaptation of male rats to repeated social encounters and experimental manipulations. *Contemp Top Lab Anim Sci* 44(2): 28–31.

Singletary, SJ. et al. 2005. Lack of correlation of vaginal impedance measurements with hormone levels in the rat. *Contemp Top Lab Anim Sci* 44(6): 37–42.

Smits, BM. et al. 2004. Genetic variation in coding regions between and within commonly used inbred rat strains. *Genome Res* 14(7): 1285–1290.

Sofikitis, N. et al. 1992. The role of the seminal vesicles and coagulating glands in fertilization in the rat. *Int J Androl* 15(1): 54–56.

Stahl, WR. 1966. Scaling of respiratory variables in mammals. *J Appl Physiol* 22(3): 453–460.

Standford, SC. et al. 1988. Deficits in exploratory behaviour in socially isolated rats are not accompanied by changes in cerebral cortical adrenoceptor binding. *J Affect Disord* 15(2): 175–180.

Stender, RN. et al. 2007. Establishment of blood analyte intervals for laboratory mice and rats by use of a portable clinical analyzer. *J Am Assoc Lab Anim Sci* 46(3): 47–52.

Stringer, SK, Seligmann, BE. 1996. Effects of two injectable anesthetic agents on coagulation assays in the rat. *Lab Anim Sci* 45(4): 430–433.

Suzuki, K. et al. 2000. Changes in plasma arginine vasopressin concentration during lactation in rats. *Comp Med* 50(3): 277–280.

Takahashi, S. et al. 2011. Comparison of the blood coagulation profiles of ferrets and rats. *J Vet Med Sci* 73(7): 953–956.

Thomas, MA. et al. 2002. Phylogenetics of rat inbred strains. *Mamm Genome* 14: 61–64.

Tisher, CC, Madsen, KM. 2000. Anatomy of the kidney. In: Brenner BM, editor. *Brenner and Rector's The Kidney*. Philadelphia, PA: WB Saunders.

Trevisan, A. et al. 1999. Glutamine synthetase activity in rat urine as sensitive marker to detect S3 segment-specific injury of proximal tubule induced by xenobiotics. *Arch Toxicol* 73(4-5): 255–262.

Tsujio, M. et al. 2008. Skin morphology of thyroidectomized rats. *Vet Pathol* 45(4): 505–511.

Turner, CH. et al. 2001. Variability in skeletal mass, structure, and biomechanical properties among inbred strains of rats. *J Bone Miner Res* 16(8): 1532–1539.

Turner, JG. et al. 2005. Hearing in laboratory animals: Strain differences and nonauditory effects of noise. *Comp Med* 55(1): 12–23.

Ulich, TR, del Castillo, J. 1991. The hematopoietic and mature blood cells of the rat: Their morphology and the kinetics of circulating leukocytes in control rats. *Exp Hematol* 19(7): 639–648.

Vachon, P, Moreau, JP. 2001. Serum corticosterone and blood glucose in rats after two jugular vein blood sampling methods: Comparison of the stress response. *Contemp Top Lab Anim Sci* 40(5): 22–24.

Vachon, P. et al. 2001. Postnatal development of penile NADPH diaphorase in male rats (*Rattus norvegicus*): An indicator of erectile function. *Contemp Top Lab Anim Sci* 40(4): 41–43.

Van Herck, H. et al. 2001. Blood sampling from the retro-orbital plexus, the saphenous vein and the tail vein in rats: Comparative effects on selected behavioural and blood variables. *Lab Anim* 35(2): 131–139.

Varga, F. 1976. Transit time changes with age in the gastrointestinal tract of the rat. *Digestion* 14(4): 319–324.

Vigueras, RM. et al. 1999. Histological characteristics of the intestinal mucosa of the rat during the first year of life. *Lab Anim* 33(4): 393–400.

Wakuta, M. et al. 2007. Delayed wound closure and phenotypic changes in corneal epithelium of the spontaneously diabetic Goto-Kakizaki rat. *Invest Ophthalmol Vis Sci* 48(2): 590–596.

Weinstein, AM. 1998. A mathematical model of the inner medullary collecting duct of the rat: Pathways for Na and K transport. *Am J Physiol* 274(5 Pt 2): F841–F855.

Weiss, DJ, Wardop, KJ. 2010. *Schalm's Veterinary Hematology*, 6th ed. Ames (IA): Blackwell Publishing.

Weiss, H. et al. 2005. Genetic analysis of the LEW.1AR1-iddm rat: An animal model for spontaneous diabetes mellitus. *Mamm Genome* 16(6): 432–441.

Wilkinson, JM. et al. 2000. Comparison of male reproductive parameters in three rat strains: Dark Agouti, Sprague-Dawley and Wistar. *Lab Anim* 34: 70–75.

Willott, JF. 2007. Factors affecting hearing in mice, rats, and other laboratory animals. *J Am Assoc Lab Anim Sci* 46(1): 23–27.

Wise, GE. 2009. Cellular and molecular basis of tooth eruption. *Orthod Craniofac Res* 12(2): 67–73.

Wolthers, T. et al. 1994. Dose-dependent stimulation of insulin-like growth factor-binding protein-1 by lanreotide, a somatostatin analog. *J Clin Endocrinol Metab* 78(1): 141–144.

Wu, WJ. et al. 2001. Aminoglycoside ototoxicity in adult CBA, C57BL and BALB mice and the Sprague-Dawley rat. *Hear Res* 158(1–2): 165–178.

Yener, T. et al. 2007 Determination of oestrous cycle of the rats by direct examination: How reliable? *Anat Histol Embryol* 36(1): 75–77.

Yokoyama, E. 1983. Ventilatory functions of normal rats of different ages. *Comp Biochem Physiol A* 75(1): 77–80.

Young, TH. et al. 2008. Quantitative rat liver function test by galactose single point method. *Lab Anim* 42(4): 495–504.

chapter 2

Animal Welfare Information Center. 1995. *Environmental Enrichment Resources for Laboratory Animals: 1965–1995.* USDA.

Arndt, SS. et al. 2010. Co-species housing in mice and rats: Effects on physiological and behavioral stress responsivity. *Horm and Behav* 57(3): 342–351.

Baker, HJ. et al. 1979. *The Laboratory Rat*, vols. 1 and 2. New York: Academic Press.

Bernard, RA, Halpern, BP. 1968. Taste changes in vitamin A deficiency. *J Gen Physiol* 52(3): 444–464.

Berrocal, Y. et al. 2007. Social and environmental enrichment improves sensory and motor recovery after severe contusive spinal cord injury in the rat. *J Neurotrauma* 24: 1761–1772.

Beys, E. et al. 1995. Ovarian changes in Sprague-Dawley rats produced by nocturnal exposure to low intensity light. *Lab Anim* 29: 335–338.

Baumans, V. 2005. Environmental enrichment for laboratory rodents and rabbits: Requirements of rodents, rabbits, and research. *ILAR J* 46(2): 162–170.

Broderson, JR. et al. 1976. The role of environmental ammonia in respiratory mycoplasmosis of rats. *Amer J Pathol* 85: 115–130.

Brown, PR, Trent, CM. 2008. Building sustainability, reducing your facilities' "carbon footprint." Presented at the 59th AALAS National Meeting, Indianapolis, IN, 9–13 Nov 2008.

Burke, DA. et al. 2007. Use of environmentally enriched housing for rats with spinal cord injury: The need for standardization. *J Am Assoc Lab Anim Sci* 46(2): 34–41.

Burn, CC. et al. 2008. Marked for life? Effects of early cage-cleaning frequency, delivery batch, and identification tail-marking on rat anxiety profiles. *Dev Psychobiol* 50: 266–277.

Centers for Disease Control and Prevention. 2009. *Biosafety in Microbiological and Biomedical Laboratories*, 5th ed.

Chmiel, DJ, Noonan, M. 1996. Preference of laboratory rats for potentially enriching stimulus objects. *Lab Anim* 30: 97–101.

Code of Federal Regulations. 2008. Animal Welfare. Federal Register Title 9 Subchapter A.

Committee on National Earthquake Resilience. et al. 2011. *National Earthquake Resilience: Research, Implementation, and Outreach.* Washington, D.C.: National Academies Press.

Cover, CE, Barron, MJ. 1998. A new feeder for diet optimization in rats. *Contemp Top Lab Anim Sci* 37(6): 84–86.

Curry, G. et al. 1998. Advances in cubicle design using computational fluid dynamics as a design tool. *Lab Anim* 32(2): 117–127.

Dahlin, J. et al. Body weight and faecal corticosterone metabolite excretion in male Sprague-Dawley rats following short transportation and transfer from group-housing to single-housing. *Scand J Lab Anim Sci* 36(2): 205–213.

Davey, AK. et al. 2003. Decrease in hepatic drug-metabolizing enzyme activities after removal of rats from pine bedding. *Comp Med* 53(3): 299–302.

Duffy, PH. et al. 2008. Nonneoplastic pathology in male Sprague-Dawley rats fed the American Institute of Nutrition 93M purified diet at ad libitum and dietary-restricted intakes. *Nutr Res* 28(3): 179–189.

Duffy, PH. et al. 2008. Neoplastic pathology in male Sprague-Dawley rats fed AIN-93M diet ad libitum or at restricted intakes. *Nutr Res* 28(1): 36–42.

Duffy, PH. et al. 2004. The effects of different levels of dietary restriction on neoplastic pathology in the male Sprague-Dawley rat. *Aging Clin Exp Res* 16(6): 448–456.

Duffy, PH. et al. 2001. The effects of different levels of dietary restriction on aging and survival in the Sprague-Dawley Rat: Implications for chronic studies. *Aging Clin Exp Res* 13(4): 263–272.

Duffy, PH. et al. 1997. The physiologic, neurologic, and behavioral effects of caloric restriction related to aging, disease, and environmental factors. *Environmental Research* 73(1–2): 242–248.

Environmental Protection Agency. 1979. Proposed Health Effects Test Standards for Toxic Substances Control Act. Federal Register, Part 2, 27334-37375, Part 4, 44054–44093.

Feirer, MR. et al. 2011. Hematologic and serum cytokine values in 2 strains of laboratory mice shipped to high altitude. Abstracts presented at the 62nd AALAS National Meeting, San Diego, CA, 2–6 Oct 2011. *J Am Assoc Lab Anim Sci* 50(5): 777.

Food and Drug Administration. 1978. Nonclinical Laboratory Studies. Good Laboratory Practice Regulations. Federal Register, Part 2, 59986–60025.

Fox, JG. et al. 2002. *Laboratory Animal Medicine*, 2nd ed. San Diego (CA): Academic Press.

Fox, JG, Cohen, BJ, Lowe, FM. 1984. *Laboratory Animal Medicine*. Orlando (FL): Academic Press.

Fernandez, CI. et al. 2004. Environmental enrichment–behavior–oxidative stress interactions in the aged rat: Issues for a therapeutic approach in human aging. *Ann NY Acad Sci* 1019: 53–57.

Fredenburg, N. et al. 2009. Effects of alternate water sources on weight gain, blood chemistries, and food consumption in Sprague Dawley rats and ICR mice following ground transportation. Abstracts presented at the 60th AALAS National Meeting, Denver, CO, 8–12 Nov 2009. *J Am Assoc Lab Anim Sci* 48(5): 602.

Galef, BC. 1981. Development of flavor preference in man and animals: The role of social and nonsocial factors. In: Aslin, RN, editor. *Development of Perception: Audition, Somatic Perception and the Chemical Senses* Volume I. New York: Academic Press.

Giral, M. et al. 2011. Effects of wire-bottom caging on heart rate, activity and body temperature in telemetry-implanted rats. *Lab Anim* 45(4): 247–253.

Gonzalez, D. et al. 2011. Failure and life cycle evaluation of watering valves. *J Am Assoc Lab Anim Sci* 50(5): 713–718.

Gonzalez, D. et al. 2008. Watering valve life cycle evaluation. Abstracts presented at the 59th AALAS National Meeting, Indianapolis, IN, 9–13 Nov 2008. *J Am Assoc Lab Anim Sci* 47(5): 120.

Guilarte, TR. et al. 2002. Environmental enrichment reverses cognitive and molecular deficits induced by developmental lead exposure. *Ann Neurol* 53(1): 50–56.

Hankenson, FC. et al. 2011. Investigation into the integrity of disposable nitrile gloves spayed with Clidox-S® in rodent facilities. Abstracts presented at the 62nd AALAS National Meeting, San Diego, CA, 2–6 Oct 2011. *J Am Assoc Lab Anim Sci* 50(5): 767.

Hart, RW. et al. 1999. Adaptive role of caloric intake on the degenerative disease processes. *Toxicol Sci* 52 (2 Suppl): 3–12.

Heeke, DS. et al. 1999. Light-emitting diodes and cool white fluorescent light similarly suppress pineal gland melatonin and maintain retinal function and morphology in the rat. *Lab Anim Sci* 43(3): 297–304.

Horne, D, Saunders, K. 2011. Auto water: An easy way to deliver medicated water to cages without the hassle of switching to water bottles. Abstracts presented at the 62nd AALAS National Meeting, San Diego, CA, 2–6 Oct 2011. *J Am Assoc Lab Anim Sci* 50(5): 764.

Howdeshell, KL. et al. 2003. Bisphenol A is released from used polycarbonate animal cages into water at room temperature. *Environ Health Perspect* 111(9): 1180–1187.

Hua J. et al. 2009. To determine the effectiveness of rice husks as bedding material for mice. SALAS National Regional Meeting Poster.

Huerkamp, MJ, Dowdy, MR. 2008. Diet replenishment for ad-libitum–fed mice housed in social groups is compatible with shelf life. *J Am Assoc Lab Anim Sci* 47(3): 47–50.

Hughes, HC, Reynolds, S. 1997. The influence of position and orientation of racks on airflow dynamics in a small animal room. *Contemp TopLab Anim Sci* 36(5): 62–67.

Hunt, PA. et al. 2009. The bisphenol A experience: A primer for the analysis of environmental effects on mammalian reproduction. *Biol Reprod* 81: 807–813.

Hunt, PA. et al. 2003. Bisphenol A exposure causes meiotic aneuploidy in the female mouse. *Curr Biol* 13: 546–553.

Institute for Laboratory Animal Research. 2010. *Disaster Planning and Management*. 51: 2.

International Air Transport Association. 2011. IATA Live Animal Regulations, 38th ed. Montreal (Quebec).

International Council for Laboratory Animal Science. 1987. *ICLAS Guidelines on the Selection and Formulation of Diets for Animals in Biomedical Research*. London, International Council for Laboratory Animal Science.

Irwin, MR. et al. 1989. Individual behavioral and neuroendocrine differences in responsiveness to audiogenic stress. *Pharmacol Biochem Behav* 32: 913–917.

Ivanisevic-Milovanovic, OK. et al. 1995. The effect of constant light on the concentration of catecholamines of the hypothalamus and adrenal glands, circulatory adrenocorticotropin hormone and progesterone. *J Endocrinol Invest* 18: 378–383.

Jacobs, WW. et al. 1978. Progress in animal flavor research. In: Bullard, R. *Flavor Chemistry of Animal Foods*. ACS Symposium Series, American Chemical Society, Washington, D.C., pp. 1–20.

Karaca, F. et al. 2003. Effects of protein deficiency on testosterone levels, semen quality, and testicular histology in the developing rat. *Scand J Lab Anim Sci* 30(1): 7–13.

Kasanen, IHE. et al. 2011. Comparison of ear tattoo, ear notching and microtattoo in rats undergoing cardiovascular telemetry. *Lab Anim* 45(4): 154–159.

Kennedy, BW, Beal, TS. 1991. Minerals leached into drinking water from rubber stoppers. *Lab Anim Sci* 43(3): 233–236.

Koehler, KE. et al. 2003. When disaster strikes: Rethinking caging materials. *Lab Anim (NY)* 32(4): 24–27.

Laboratory Animal Breeders Association of Great Britain Limited/ Laboratory Animal Science Association. 1993. Guidelines for the care of laboratory animals in transit. *Lab Anim* 27: 93–107.

Latalladi, T. et al. 2010. Novel technique for handling, sterilizing, and storage of removable water valves used with an automatic watering system. Abstracts presented at the 61st AALAS National Meeting, Atlanta, GA, 10–14 Oct 2010. *J Am Assoc Lab Anim Sci* 49(5): 715.

Lawson, DL. et al. 2000. The effects of housing enrichment on cardiovascular parameters in spontaneously hypertensive rats. *Contemp Top Lab Anim Sci* 39(1): 9–13.

Le, N-MP, Brown, JW. 2008. Characterization of the thermoneutral zone of the laboratory rat. 2008 *FASEB Meeting Abstracts* 22: 956.19.

Leakey, JEA. et al. 1994. Role of glucocorticoids and "caloric stress" in modulating the effects of caloric restriction in rodents. *Ann NY Acad Sci* 719: 171–194.

Lewis, SM. et al. 2003. Nutrient intake and growth characteristics of male Sprague-Dawley Rats fed AIN-93M purified diet or NIH-31 natural-ingredient diet in a chronic two-year study. *Aging Clin Exp Res* 15(6): 460–468.

Lipman, NS. 1993. Strategies for architectural integration of ventilated caging systems. *Contemp Top Lab Anim Sci* 32(1): 7–10.

Lohmiller, J, Lipman, NS. 1998. Silicon crystals in water of autoclaved glass bottles. *Contemp Top Lab Anim Sci* 7(1): 62–65.

Mani SK. et al. 2005. Disruption of male sexual behavior in rats by tetrahydrofurandiols (THF-diols). *Steroids* 70: 750–775.

Manser, CE. et al. 1996. The use of a novel operant test to determine the strength of preference for flooring in laboratory rats. *Lab Anim* 30: 1–6.

Manser, CE. et al. 1995. An investigation into the effects of cage flooring on the welfare of laboratory rats. *Lab Anim* 29: 353–363.

Markaverich, BM. et al. 2007. Tetrahydrofurandiols (THF-diols), leukotoxindiols (LTX-diols), and endocrine disruption in rats. *Environ Health Perspect* 115: 702–708.

Markaverich, BM. et al. 2005. Leukotoxin diols from ground corncob bedding disrupt estrous cyclicity in rats and stimulate MCF-7 breast cancer cell proliferation. *Environ Health Perspect* 113: 1698–1704.

Markaverich, BM. et al. 2002. A novel endocrine disrupting agent in corn with mitogenic activity in human breast and prostatic cancer cells. *Environ Health Perspect* 110: 169–777.

McLennan, IS, Taylor-Jeffs, J. 2004. The use of sodium lamps to brightly illuminate mouse houses during their dark phases. *Lab Anim* 38: 384–392.

Meijer, MK. et al. 2009. There's a rat in my room! Now what? Mice show no chronic physiological response to the presence of rats. *J Appl Anim Welf Sci* 12(4): 293–305.

Muhlhauser, A. et al. 2009. Bisphenol A effects on the growing mouse oocyte are influenced by diet. *Biol Reprod* 80: 1066–1071.

Müller, G. et al. 1992. A new device for long-term continuous enteral nutrition of rats by elementary diet via gastrostomy, following extensive oesophageal or lower gastrointestinal surgery. *Lab Anim* 26: 9–14.

Myers, JP. et al. 2009. Why public health agencies cannot depend on good laboratory practices as a criterion for selecting data: The case of Bisphenol A. *Environ Health Perspect* 117(3): 309–315.

Nagamine, CM. et al. 2011. Development of hypercapnic and hypoxic conditions in disposable individually ventilated cages following removal from mechanical ventilation. Presented at the 62nd AALAS National Meeting, San Diego, CA, 2–6 Nov 2011.

National Research Council. 2010. *Guide for the Care and Use of Laboratory Animals.* Washington, D.C.: National Academy Press.

National Research Council. 2006. *Guidelines for the Humane Transportation of Research Animals.* Washington, D.C.: National Academy Press.

National Research Council. 1996. *Rodents.* Washington, D.C.: National Academy Press.

National Research Council. 1995. *Nutrient Requirements of Laboratory Animals,* 4th revised ed. Washington, D.C.: National Academy Press.

National Research Council. 1989. *Biosafety in the Laboratory: Prudent Practices for the Handling and Disposal of Infectious Materials.* Washington, D.C.: National Academy Press.

Nayfield, KC, Besch, EL. 1981. Comparative responses to rabbits and rats to elevated noise. *Lab Anim Sci* 34(4): 386–390.

Norton, JN. et al. 2011. Comparative vibration levels perceived among species in a laboratory animal facility. *J Am Assoc Lab Anim Sci* 50(5): 653–659.

Patterson-Kane, EG. 2003. Shelter enrichment for rats. *Contemp Top Lab Anim Sci* 42(2): 46–48.

Peace, TA. et al. 2001. Effects of caging type and animal source on the development of foot lesions in Sprague Dawley rats (*Rattus norvegicus*). *Contemp Top Lab Anim Sci* 40(5): 17–21.

Perkins, S.E., Lipman, N. S. 1995. Characterization and quantification of microenvironmental contaminants in isolator cages with a variety of contact beddings. *Contemp Top Lab Anim Sci* 34(3): 93–98.

Pfaff, J, Stecker, M. 1976. Loudness level and frequency content of noise in the animal house. *Lab Anim* 10: 111–117.

Phoenix Controls Corporation. 2002. Cage rack ventilation options for laboratory animal facilities: White Paper. Phoenix Controls Corporation, Newton, Massachusetts.

Potgieter, FJ, Wilke, PI. 1996. The dust content, dust generation, ammonia production, and absorption properties of three different rodent bedding types. *Lab Anim* 30: 79–87.

Pritchett-Corning, KR. et al. 2009. Breeding and housing laboratory rats and mice in the same room does not affect the growth or reproduction of either species. *J Am Assoc Lab Anim Sci* 48(5): 492–498.

Public Health Service. 2002. *Public Health Service Policy on Humane Care and Use of Laboratory Animals.* Washington, D.C.: U.S. Department of Health and Human Services.

Pulgar, R. et al. 2000. Determination of Bisphenol A and related aromatic compounds released from Bis-GMA-based composites and sealants by high performance liquid chromatography. *Environ Health Perspect* 108(1): 21–27.

Rao, GN, Knapka, JJ. 1987. Contaminant and nutrient concentrations of natural ingredient rat and mouse diet used in chemical toxicology studies. *Fundam Appl Toxicol* 9: 329–338.

Rosa, MA. et al. 2009. Quality control of the automated watering system at IDIBELL. SECAL Conference Poster.

Refinetti, R, Horvath, SM. 1989. Thermopreferendum of the rat: Inter- and intra-subject variabilities. *Behav Neural Biol* 52(1): 87–94.

Reynolds, RP. et al. 2010. Noise in a laboratory animal facility from the human and mouse perspectives. *J Am Assoc Lab Anim Sci* 49(5): 592–597.

Roberts, SB, Coward, WA. 1985. The effects of lactation on the relationship between metabolic rate and ambient temperature in the rat. *Ann Nutr Metab* 29(1): 19–22.

Rock, FM. et al. 1997. Effects of caging type and group size on selected physiologic variables in rats. *Contemp Top Lab Anim Sci* 36(2): 69–72.

Romanovsky, AA. et al. 2002. Selected contribution: Ambient temperature for experiments in rats: A new method for determining the zone of thermal neutrality. *J Appl Physiol* 92: 2667–2679.

Rozema, R. 2011. Acoustic and vibration levels in research animal facilities: What is required? *ALN* 10(3): 25–29.

Rozema, R. 2009. Noise and vibration considerations for the animal lab environment. *ALN* 8(3): 23–31.

Ruys, T. 1991. *The Handbook of Facilities Planning,* Volume 2: Laboratory Animal Facilities. New York: Van Nostrand Reinhold.

Saibaba, P. et al. 1995. Behaviour of rats in their home cages: Daytime variations and effects of routine husbandry procedures analysed by time sampling techniques. *Lab Anim* 30(1): 13–21.

Schondelmeyer, CW. et al. 2006. Investigation of appropriate sanitization frequency for rodent caging accessories: Evidence supporting less-frequent cleaning. *J Am Assoc Lab Anim Sci* 45(6): 40–43.

Sebesteny, A. et al. 1992. Microbiologically monitored fumigation of a newly built SPF laboratory rodent facility. *Lab Anim* 26: 132–139.

Sharp, GP. 2010. Demand-based control of lab air change rates. *ASHRAE Journal* 52(2): 30–41.

Sharp, GP. 2009 A comprehensive review of the IEQ and energy savings impact of dynamically varying air change rates in labs and vivariums. *ALN* 8(2): 21–27.

Sharp, GP. 2008. Dynamic variation of laboratory air change rates: A new approach to saving energy and enhancing safety. *ALN* 7(8): 11–15.

Sharp, J. et al. 2005. Effects of a cage enrichment program on heart rate, blood pressure, and activity of male Sprague-Dawley and Spontaneously Hypertensive rats monitored by radiotelemetry. *Contemp Top Lab Anim Sci* 44(2): 32–40.

Simpson, J, Kelly, JP. 2011 The impact of environmental enrichment in laboratory rats: Behavioural and neurochemical aspects. *Behav Brain Res* 222: 246–264.

Smith, DJA, Hughes, GS. 2007. Sodium lighting as a viable alternative to red lighting within the vivarium. Abstracts presented at the 58th AALAS National Meeting, Charlotte, NC, 14–18 Oct 2007. *J Am Assoc Lab Anim Sci* 46(4): 112.

Smith, V, Coleman, M. 2010. Life with automatic water. IAT (UK) Conference Poster.

Stakutis, RE. 2003. Cage rack ventilation options for laboratory animal facilities. *Lab Anim* 32(8): 47–52.

Suckow, MA. et al. 2005. *The Laboratory Rat*, 2nd ed. Burlington (MA): Academic Press.

Susiarjo, M. et al. 2007. Bisphenol A exposure in utero disrupts early oogenesis in the mouse. *PLoS Genet* Jan 12,3(1): 63–70.

Thomas, ML. et al. 2001. The effect of magnesium deficiency on volatile anaesthetic requirement in the rat: The role of central noradrenergic neuronal activity. *Magnes Res* 14(3): 195–201.

Tinwell, H. et al. 2002. Normal sexual development of two strains of rat exposed in utero to low doses of bisphenol A. *Toxicol Sci* 68: 339–348.

Tober-Meyer, BK, Bieniek, HJ. 1981. Studies on the hygiene of drinking water for laboratory animals. 1. The effect of various treatments on bacterial contamination. *Lab Anim* 15: 107–110.

Tober-Meyer, BK. et al. 1981. Studies on the hygiene of drinking water for laboratory animals. 2. Clinical and biochemical studies in rats and rabbits during long-term provision of acidified water. *Lab Anim* 15: 111–117.

Traknyak, F. 2006. Cut costs by controlling airflow. *Engineering Systems* 23(9): 52.

van der Harst, JE. et al. 2003. Standard housed rats are more sensitive to rewards than enriched housed rats as reflected by their anticipatory behaviour. *Behav Brain Res* 142: 151–156.

van Loo, PLP, Baumans, V. 2004. The importance of learning young: The use of nesting material in laboratory rats. *Lab Anim* 38: 17–24.

van Ruiven, R. et al. 1996. Adaption period of laboratory animals after transport: A review. *Scand J Lab Anim Sci* 23(4): 85–190.

Vento, PJ. et al. 2008. Food intake in laboratory rats provided standard and fenbendazole-supplemented diet. *J Am Assoc Lab Anim Sci* 47(6): 46–50.

Verce, MF, 2008. Minimizing decomposition of vaporized hydrogen peroxide for biological decontamination of galvanized steel ducting. *Environ Sci Technol* 42: 5765–5771.

Voipio, H. 1997. How do rats react to sound? *Scand J Lab Anim Sci* 24 (Supplement 1).

vom Saal, FS. et al. 2010. Flawed experimental design reveals the need for guidelines requiring appropriate positive controls in endocrine disruption research. *Toxicol Sci* 115(2): 612–613.

Waldrab, D. 2009. Hold the steam. *ALN* 8(8): 21–23.

Waynforth, HB, Flecknell, PA. 1992. *Experimental and Surgical Technique in the Rat*. London (UK): Academic Press.

Ward, GM. et al. 2009. Humidity and cage and bedding temperatures in unoccupied static mouse caging after steam sterilization. *J Am Assoc Lab Anim Sci* 48(6): 774–779.

Wingfield, WE, Palmer, SB (eds). 2009. *Veterinary Disaster Response*. Ames (IA): Wiley-Blackwell.

Witt, S. et al. 2011. Nylon cylinder: Safe and cost-effective chewable rat enrichment. Abstracts presented at the 62nd AALAS National Meeting, San Diego, CA, 2–6 Oct 2011. *J Am Assoc Lab Anim Sci* 50(5): 782.

Wolf, NS. et al. 2000. Normal mouse and rat strains as models for age-related cataract and the effect of caloric restriction on its development. *Exp Eye Res* 70(5): 683–692.

Zondek, B, Tamari, I. 1960. Effect of audiogenic stimulation on genital function and reproduction. *Am J Obstet Gynecol* 80(6): 1041–1048.

Zutphen, L.F.M. et al. 1993. *Principles of Laboratory Animal Science: A Contribution to the Humane Use and Care of Animals and to the Quality of Experimental Results.* Amsterdam: Elsevier Science Publishers.

---

chapter 3

AAALAC International. [Internet]. Categories of Accreditation [Cited 21 February 2012] Available at: www.aaalac.org/accreditation/categories.cfm

Austria. [Internet]. Tierschutzgesetz (Animal Protection Act) [Cited 27 September 2011] Available at: www.ris.bka.gv.at/Dokumente/Erv/ERV_2004_1_118/ERV_2004_1_118.pdf

Bayne, K. et al. 2010. Legislation and oversight of the conduct of research using animals: A global overview. In: Hubrecht, R, Kirkwood, J, editors. *UFAW Handbook*, 8th ed. Oxford (UK): Wiley-Blackwell, pp. 107–123.

Bureau of Labor Statistics, U.S. Department of Labor. [Internet] Table 10. Number, percent, and incidence rate of nonfatal occupational injuries and illnesses involving days away from work by selected worker and case characteristics and musculoskeletal disorders, all United States, private industry, 2008. [Cited 19 October 2011] Available at: www.bls.gov/iif/oshwc/osh/case/ostb2211.pdf

Croatia. [Internet]. (Decision Promulgating) The Animal Protection Act [Cited 27 September 2011] Available at: www.prijatelji-zivotinja.hr/index.en.php?id=470

Cullinan, P. et al. 1999. Allergen exposure, atopy and smoking as determinants of allergy to rats in a cohort of laboratory employees. *Eur Respir J* 13: 1139–1143.

Chile Ministerio de Salud. [Internet] Ley Número 20.380 Sobre Protección de Animales. [Cited 27 September 2011] Available at: www.fao.org/fileadmin/user_upload/animalwelfare/LEY-20380_03-OCT-2009_1.pdf

Code of Federal Regulation (US, CFR). 1992. 1910.1030 Bloodborne Pathogens. US Department of Labor, Occupational Safety and Health Administration.

Cole, JS. et al. 2005. *Ornithonyssus bacoti* infestation and eradication from a mouse colony. *Contemp Top Lab Anim Sci* 44(5): 27–30.

Cost Rica. [Internet]. La Gaceta Diario Oficial, Ley de Promocion del Desarrollo Cientifico y Tecnologico Numero 7169 del 26 de Junio de 1990, la Ley de Bienestar de los Animales Numero 7451 del 13 de Deciembre de 1994. [Cited 27 September 2011] Available at: www.glin.gov/view.action?glinID=59264

Council for International Organizations of Medical Sciences (CIOMS). 1985. *International Guiding Principles for Biomedical Research Involving Animals.*

Earle, CD. et al. 2007. High-throughput fluorescent multiplex array for indoor allergen exposure assessment. *J Allergy Clin Immunol* 119(2): 428–433.

Eggleston, PA. et al. 1990. Occupational challenge studies with laboratory workers allergic to rats. *J Allergy Clin Immunol* 86: 63–72.

Eggleston, PA. et al. 1989. Task-related variation in airborne concentrations of laboratory animal allergens: Studies with rat N 1. *J Allergy Clin Immunol* 84: 347–352.

Elliott, SP. 2007. Rat bite fever and *Streptobacillus moniliformis*. *Clin Microbiol Rev* 20(1): 13–22.

Estonia [Internet]. Animal Protection Act [Cited 27 September 2011] Available at: www.legaltext.ee/text/en/X60004K4.htm

Finland Ministry of Agriculture and Forestry. [Internet]. Animal Welfare Act. [Cited 26 September 2011]. Available at: www.finlex. fi/fi/laki/kaannokset/1996/en19960247.pdf

Gao, C. Current laboratory animal laws regulations, policies, and administration in China. Presented at Shanghai Sharing Conference 2010.

Hill, WA. et al. 2005. Use of permethrin eradicated the tropical rat mite (*Ornithonyssus bacoti*) from a colony of mutagenized and transgenic mice. *Contemp Top Lab Anim Sci* 44(5): 31–34.

Home Office. [Internet]. Animals Scientific Procedures Inspectorate and Division Annual Report 2010. [Cited 12 March 2012]. Available at: www.homeoffice.gov.uk/publications/science-research-statistics/ animals/annual-reports/animals-annual-report-2010?view=Binary

India Ministry of Environment and Forests. [Internet]. The Animal Welfare Act, 2011. [Cited Cited 26 September 2011]. Available at: http://moef.nic.in/downloads/public-information/draft-animal-welfare-act-2011.pdf

International Ergonomics Association. [Internet]. Definition of Ergonomics, The Discipline of Ergonomics. [Cited 19 October 2011] Available at: http://iea.cc/01_what/What%20is%20Ergonomics.html

Jeal, H. et al. 2009. Dual sensitization to rat and mouse urinary allergens reflects cross-reactive molecules rather than atopy. *Allergy* 64: 855–861.

Jeal, H. et al. 2004. Determination of the T Cell epitopes of the lipocalin allergen, Rat n 1. *Clin Exp Allergy* 34: 1919–1925.

Kenya Law Reports. [Internet]. (Chapter 360) Prevention of Cruelty to Animals Act. [Cited Cited 26 September 2011]. Available at: www.kenyalaw.org/kenyalaw/klr_app/frames.php

Kimwele, C. et al. 2011. A Kenyan perspective on the use of animals in science education and research in Africa and prospects for improvement. *Pan Afr Med J* 9: 45.

Koets, M. et al. 2011. Rapid one-step assays for on-site monitoring of mouse and rat allergens. *J Environ Monit* 13: 3475–3480.

Krop, EJM. et al. 2007. Spreading of occupational allergens: Laboratory animal allergens on hair-covering caps and in mattress dust of laboratory animal workers. *Occup Environ Med* 64: 267–272.

Lieutier-Colas, F. et al. 2002. Prevalence of symptoms, sensitization to rats, and airborne exposure to major rat allergen (Rat n 1) and to endotoxin in rat-exposed workers: A cross-sectional study. *Clin Exp Allergy* 32: 1424–1429.

Malaysia. [Internet]. Laws of Malaysia Act 647 Animals Act 1953, The Revision of Laws (Rectification of Animals Act 1953) Order 2006 [Cited 19 October 2011] Available at: www.dvs.gov.my/c/document_library/get_file?uuid=feec289c-db0d-4679-bd04-41c52e7e8add&groupId=28711

McLeod, V, Ketcham, G. 2008. Needling you can handle: Bloodborne pathogens—Part 1. *ALN* 7(6): 51–52.

McLeod, V, Ketcham, G. 2008. Stuck on you: Bloodborne pathogens—Part 2. *ALN* 7(7): 65–67.

New Zealand Ministry of Agriculture and Forestry. [Internet]. The Animal Welfare Act—A Framework for the 21st Century. [Cited 23 October 2011]. Available at: www.biosecurity.govt.nz/legislation/animal-welfare-act/index.htm

Norway Information from the Government and the Ministries. [Internet]. Animal Welfare Act [Cited 26 September 2011]. Available at: www.regjeringen.no/en/doc/laws/Acts/animal-welfare-act.html?id=571188

National Research Council. 2010. *Guide for the Care and Use of Laboratory Animals.* Washington, D.C.: National Academy Press.

Office of Laboratory Animal Welfare. [Internet]. Animal Welfare Assurance for Foreign Institutions. [Cited 24 October 2011]. Available at: grants.nih.gov/grants/olaw/sampledoc/foreign.htm

Papanicolas, LE. et al. 2012. Meningitis and pneumonitis caused by pet rodents. *Med J Aust* 196(3): 202–203.

Poland. [Internet]. Poland Animal Protection Act of August 21, 1997. [Cited 27 September 2011] Available at: www.internationalwildlifelaw.org/PolandAnimalProtectionAct.html

Reeb-Whitaker, CK. et al. 1999. Control strategies for aeroallergens in an animal facility. *J Allergy Clin Immunol* 103: 139–146.

Renström, A. et al. 2001. Working with male rodents may increase risk of allergy to laboratory animals. *Allergy* 56: 964–970.

Russell, WMS, Burch, RL. 1959. *The Principles of Humane Experimental Technique. London: Methuen. [Revised edition 2009, as The Three R's and the Humanity Criterion.]*

Saarelainen, S. et al. 2007. Animal-derived lipocalin allergens exhibit immunoglobulin E cross-reactivity. *Clin Exp Allergy* 38: 374–381.

Science Council of Japan. 2006. *Guidelines for Proper Conduct of Animal Experiments.*

Shalev, M. 2011. Regulation update. *ALN World* 4(4): 6–7.

South Africa. [Internet]. Veterinary and Para-veterinary Professions Act No. 19 of 1982. [Cited 19 April 2012] Available at: www.savc.org.za/pdf_docs/act_19_of_1982.pdf

South African Bureau of Standards. 2008. *The Care and Use of Animals for Scientific Purposes.* Pretoria.

Tanzania. [Internet]. The Animal Welfare Act, 2008. [Cited 26 September 2011]. Available at: www.fao.org/fileadmin/user_upload/animalwelfare/tanzania.pdf

Walls, AF, Longbottom, JL. 1985. Comparison of rat fur, urine, saliva, and other rat allergen extracts by skin testing, RAST, and RAST inhibition. *J Allergy Clin Immunol* 75(2): 242–251.

World Organisation for Animal Health. 2010. *Terrestrial Animal Health Code.*

Zimbabwe Law and Justice. [Internet]. Animals, Plants & Forestry. [Cited 26 September 2011]. Available at: www.law.co.zw/index. php?option=com_docman&task=cat_view&gid=47&Itemid=38

chapter 4

Abrass, CK. 2000. The nature of chronic progressive nephropathy in aging rats. *Adv Ren Replace Ther* 7(1): 4–10.

American College of Laboratory Animal Medicine. [Internet]. 2001. Position statement on rodent surgery. [Cited 23 Oct 2011]. Available at: www.aclam.org

American Veterinary Medical Association. [Internet]. 2007. AVMA guidelines on euthanasia. [Cited 23 Oct 2011]. Available at: www.avma.org

Amouzadeh, HR. et al. 1991. Xylazine-induced pulmonary edema in rats. *Toxicol Appl Pharmacol* 108(3): 417–427.

Artwohl, J. et al. 2006. Report of the ACLAM task force on rodent euthanasia. *J Am Assoc Lab Anim Sci* 45(1): 98–105.

Artwohl, JE. et al. 1994. The Efficacy of a dirty bedding sentinel system for detecting sendai virus infection in mice: A comparison of clinical signs and seroconversion. *Lab Animal Sci* 44: 73–75.

Avsaroglu, H. et al. 2007. Differences in response to anaesthetics and analgesics between inbred rat strains. *Lab Anim* 41(3): 337–344.

Balsari, A. et al. Dermatophytes in clinically healthy laboratory animals. *Lab Anim* 15(1): 75–77.

Baker, DG. 1998. Natural pathogens of laboratory mice, rats, and rabbits and their effects on research. *Clin Microbiol Rev* 11(2): 231–266.

Baker, DG. 2007. Parasites of rats and mice. In: Baker DG, editor. *Flynn's Parasites of Laboratory Animals*. Ames (IA): Blackwell Publishing, pp. 304–397.

Battles, AH. et al. 1987. Efficacy of ivermectin against natural infection of *Syphacia muris* in rats. *Lab Anim Sci* 37(6): 791–792.

Bennett, M. et al. 1997. Cowpox in British voles and mice. *J Comp Pathol* 116(1): 35–44.

Besselsen, DG. et al. 2008. Lurking in the shadows: Emerging rodent infectious diseases. *ILAR J* 49(3): 277–290.

Bihun, CG, Percy, DH. 1995. Morphologic changes in the nasal cavity associated with sialodacryoadenitis virus infection in the Wistar rat. *Vet Pathol* 32(1): 1–10.

Boehm, CA. et al. 2010. Midazolam enhances the analgesic properties of dexmedetomidine in the rat. *Vet Anaesth Analg* 37(6): 550–556.

Boksa, P. et al. 1995. Effects of a period of asphyxia during birth on spatial learning in the rat. *Pediatr Res* 37(4 Pt 1): 489–496.

Boorman, GA, Hollander, CF. 1974. High incidence of spontaneous urinary bladder and ureter tumors in the Brown Norway rat. *J Natl Cancer Inst* 52: 1005–1008.

Boot, R. et al. 1995. Serological studies of *Corynebacterium kutscheri* and coryneform bacteria using an enzyme-linked immunosorbent assay (ELISA). *Lab Anim* 29(3): 294–299.

Bootz, F. et al. 2003. Comparison of the sensitivity of in vivo antibody production tests with in vitro PCR-based methods to detect infectious contamination of biological materials. *Lab Anim* 37(4): 341–351.

Brammer, A. et al. 1993. A comparison of propofol with other injectable anaesthetics in a rat model for measuring cardiovascular parameters. *Lab Anim* 27(3): 250–257.

Brown, MJ. et al. 1993. Guidelines for animal surgery in research and teaching: AVMA panel on animal surgery in research and teaching, and the ASLAP (American Society of Laboratory Animal Practitioners). *Am J Vet Res* 54(9): 1544–1559.

Campe, H. et al. 2009. Cowpox virus transmission from pet rats to humans, Germany. *Emerg Infect Dis* 15(5): 777–780.

Cappon, GD. et al. 2003. Relationship between cyclooxygenase 1 and 2 selective inhibitors and fetal development when administered to rats and rabbits during the sensitive periods for heart development and midline closure. *Birth Defects Res B Dev Reprod Toxicol* 68(1): 47–56.

Carthew, P, Aldred, P. 1988. Embryonic death in pregnant rats owing to intercurrent infection with sendai virus and *Pastereulla pneumotropica. Lab Anim* 22(1): 92–97.

Carty, AJ. 2008. Opportunistic infections of mice and rats: Jacoby and Lindsey revisited. *ILAR J* 49(3): 272–276.

Charles River Laboratories. [Internet]. 2011. *Serologic Methods Manual: Multiplexed Fluorometric ImmunoAssay® (MFIA®).* [Cited 23 Oct 2011]. Available at: www.criver.com

Charreau, BL. et al. 1996. Transgenesis in rats: Technical aspects and models. *Transgen Res* 5(4): 223–234.

Cheong, SH. et al. 2010. Blind oral endotracheal intubation of rats using a ventilator to verify correct placement. *Lab Anim* 44(3): 278–280.

Clark, JA, Jr. et al. 1997. Pica behavior associated with buprenorphine administration in the Rat. *Lab Anim Sci* 47(3): 300–303.

Clifford, CB, Watson, J. 2008. Old enemies, still with us after all these years. *ILAR J* 49(3): 291–302.

Close, B. et al. 1996. Recommendations for euthanasia of experimental animals: Part 1. DGXI of the European Commission. *Lab Anim* 30(4): 293–316.

Close, B. et al. 1997. Recommendations for euthanasia of experimental animals: Part 2. DGXT of the European Commission. *Lab Anim* 31(1): 1–32.

Cohen, JK. et al. 2007. Pancreaticoduodenal arterial rupture and hemoabdomen in ACI/SegHsd rats with polyarteritis nodosa. *Comp Med* 57(4): 370–376.

Colpaert, FC. et al. 1980. Self-administration of the analgesic suprofen in arthritic rats: Evidence of *Mycobacterium butyricum*-induced arthritis as an experimental model of chronic pain. *Life Sci* 27(11): 921–928.

Compton, SR. et al. 1999. Comparison of the pathogenicity in rats of rat coronaviruses of different neutralization groups. *Lab Anim Sci* 49(5): 514–518.

Compton, SR. et al. 2004. Efficacy of three microbiological monitoring methods in a ventilated cage rack. *Comp Med* 54(4): 382–392.

Compton, SR. et al. 2004. Microbiological monitoring in individually ventilated cage systems. *Lab Anim (NY)* 33(10): 36–41.

Cook, JC. et al. 2003. Analysis of the nonsteroidal anti-inflammatory drug literature for potential developmental toxicity in rats and rabbits. *Birth Defects Res B Dev Reprod Toxicol* 68(1): 5–26.

Coons, O. 2005. P35 Infestation of spiny rat mites in vivarium. Abstract. *Contemp Top Lab Anim* 44: 79.

Cooper, DM. et al. 2000. The thin blue line: A review and discussion of aseptic technique and postprocedural infections in rodents. *Contemp Top Lab Anim Sci* 39(6): 27–32.

Cowan, A. et al. 1977. The animal pharmacology of buprenorphine, an oripavine analgesic agent. *Br J Pharmacol* 60(4): 547–554.

Criado, AB. et al. 2000. Reduction of isoflurane MAC with buprenorphine and morphine in rats. *Lab Anim* 34(3): 252–259.

Crippa, L. et al. 2000. Ringtail in suckling Munich Wistar Fromter rats: A histopathologic study. *Comp Med* 50(5): 536–539.

Cundiff, DD. et al. 1994. Characterization of cilia-associated respiratory bacillus isolates from rats and rabbits. *Lab Anim Sci* 44(4): 305–312.

Cundiff, DD. et al. 1995. Characterization of cilia-associated respiratory bacillus in rabbits and analysis of the 16S rRNA gene sequence. *Lab Anim Sci* 45(1): 22–26.

Cundiff, DD. et al. 1995. Failure of a soiled bedding sentinel system to detect ciliary associated respiratory bacillus infection in rats. *Lab Animal Sci* 45(2): 219–221.

Cushion, MT. et al. 1993. Genetic stability and diversity of *Pneumocystis carinii* infecting rat colonies. *Infect Immun* 61: 4801–4813.

Cushion, MT. et al. 2004. Molecular and phenotypic description of *Pneumocystis wakefieldiae* sp. nov., a new species in rats. *Mycologia* 96(3): 429–438.

Danneman, PJ, Mandrell, TD. 1997. Evaluation of five agents/methods for anesthesia of neonatal rats. *Lab Anim Sci* 7(4): 386–395.

Das, A. 2011. Heat Stress-induced hepatotoxicity and its prevention by resveratrol in rats. *Toxicol Mech Methods* 21(5): 393–399.

de Mello, WF, Kirkman, E. 1992. Propofol and excitatory sequelae in rats. *Anaesthesia* 47(2): 174–175.

Del Carmen Martino-Cardona, M. et al. 2010. Eradication of *Helicobacter* spp. by using medicated diet in mice deficient in functional natural killer cells and complement factor D. *J Am Assoc Lab Anim Sci* 49(3): 294–299.

Dickinson, AL. et al. 2009. Influence of early neonatal experience on nociceptive responses and analgesic effects in rats. *Lab Anim* 43(1): 11–16.

Diesch, TJ. et al. 2009. Electroencephalographic responses to tail clamping in anaesthetized rat pups. *Lab Anim* 43(3): 224–231.

Directive 2010/63/EU. 2010. [Internet]. Directive 2010/63/EU of the European Parliament and of the Council of 22 September 2010 on the Protection of Animals Used for Scientific Purposes. [Cited 16 June 2011]. Available at: http://eur-lex.europa.eu/LexUriServ/LexUriServ.do?uri=OJ:L:2010:276:0033:0079:EN:pdf

Dix, J. et al. 2004. Assessment of methods of destruction of *Syphacia muris* eggs. *Lab Anim* 38(1): 11–16.

Donnelly, TM, Quimby, FW. 2002. Biology and diseases of other rodents. In: Fox JG, Anderson LC, Loew FM, Quimby FW, editors. *Laboratory Animal Medicine*, 2nd ed. San Diego (CA): Academic Press, pp. 121–165.

Dontas, IA. et al. 2010. Malocclusion in aging Wistar rats. *J Am Assoc Lab Anim Sci* 49(1): 22–26.

Drake, MT. et al. 2008. Differential susceptibility of SD and CD rats to a novel rat theilovirus. *Comp Med* 58(5): 458–464.

Drazenovich, NL. et al. 2002. Detection of rodent *Helicobacter* spp. by use of fluorogenic nuclease polymerase chain reaction assays. *Comp Med* 52(4): 347–353.

Drummond, JC. et al. 1996. Use of neuromuscular blocking drugs in scientific investigations involving animal subjects: The benefit of the doubt goes to the animal. *Anesthesiology* 85(4): 697–699.

Dyson, MC. 2010. Management of an outbreak of rat theilovirus. *Lab Anim* 39(5): 155–157.

Easterbrook, J. et al. 2008. A survey of rodent-borne pathogens carried by wild-caught Norway rats: A potential threat to laboratory rodent colonies. *Lab Anim* 42(1): 92–98.

Eis-Hübinger, AM. et al. 1990. Fatal cowpox-like virus infection transmitted by cat. *Lancet* 336(8719): 880.

Ellerman, KE. et al. 1996. Kilham rat triggers T-cell-dependent autoimmune diabetes in multiple strains of rat. *Diabetes* 45(5): 557–562.

Elliott, SP. 2007. Rat bite fever and *Streptobacillus moniliformis*. *Clin Microbiol Rev* 20(1): 13–22.

Esatgil, MU. et al. 2008. Use of topical selamectin for the treatment of *Syphacia muris* infection in laboratory rats. *Pol J Vet Sci* 11(1): 67–69.

Federation of European Laboratory Animal Science Associations. 1994. Recommendations for the health monitoring of mouse, rat, hamster, guinea pig, and rabbit breeding colonies. *Lab Anim* 28(1): 1–12.

Feldman, DB, Seely, JC. 1988. *Necropsy Guide: Rodents and the Rabbit.* Boca Raton (FL): CRC Press.

Feldman, SH. et al. 1996. Detection of *Pneumocystis carinii* in rats by polymerase chain reaction: Comparison of lung tissue and bronchoalveolar lavage specimens. *Lab Anim Sci* 46(6): 628–634.

Field, KJ. et al. 1993. Anaesthetic effects of chloral hydrate, pentobarbitone and urethane in adult male rats. *Lab Anim* 27(3): 258–269.

Fisher, M. et al. 2007. Efficacy and safety of selamectin (Stronghold®/ Revolution™) used off-label in exotic pets. *Intern J Appl Res Vet Med* 5(3): 87–96.

Flecknell, PA. 2009. *Laboratory Animal Anaesthesia*, 3rd ed. London (UK): Elsevier.

Flecknell, PA. et al. 1999. Use of oral buprenorphine ('buprenorphine jello') for postoperative analgesia in rats: A clinical trial. *Lab Anim* 33(2): 169–174.

Flecknell, P, Waterman-Pearson, A. 2000. *Pain Management in Animals.* London (UK): WB Saunders.

Foley, PL. et al. 2011. Evaluation of a sustained-release formulation of buprenorphine for analgesia in rats. *J Am Assoc Lab Anim Sci* 50(2): 198–204.

Fox, JG. et al. 1977. Ulcerative dermatitis in the rat. *Lab Anim Sci* 27(5 Pt1): 671–678.

Fray, MD. et al. 2008. Upgrading mouse health and welfare: direct benefits of a Large-scale rederivation programme. *Lab Anim* 42(2): 127–139.

Gaertner, DJ. et al. 2008. Anesthesia and analgesia for laboratory rodents. In: Fish RE, Danneman PJ, Brown M, Karas AZ, editors. *Anesthesia and Analgesia in Laboratory Animals*, 2nd ed. London (UK): Academic Press, pp. 239–297.

Gillingham, MB. et al. 2001. A Comparison of two opioid analgesics for relief of visceral pain induced by intestinal resection in rats. *Contemp Top Lab Anim Sci* 40(1): 21–26.

Golder, FJ. et al. 2003. Urethane Anesthesia in adult female rats: Preliminary observations. *Vet Anaesth Analg* 30(2): 115.

Goto, K. et al. 1995. Detection of cilia-associated respiratory bacillus in experimentally and naturally infected mice and rats by the polymerase chain reaction. *Exp Anim* 44(4): 333–336.

Gross, DR. et al. 2003. Critical anthropomorphic evaluation and treatment of postoperative pain in rats and mice. *J Am Vet Med Assoc* 222(11): 1505–1510.

Gwynne, BJ, Wallace, J. 1992. A modified anaesthetic induction chamber for Rats. *Lab Anim* 26(3): 163–166.

Habermann, RT, Williams, FP, Jr. 1958. The identification and control of helminths in laboratory animals. *J Natl Cancer Inst* 20(5): 979–1009.

Hacker, SO. et al. 2005. A comparison of target–controlled infusion versus volatile inhalant anesthesia for heart rate, respiratory rate, and recovery time in a rat model. *Contemp Top Lab Anim Sci* 44(5): 7–12.

Haines, DC. et al. 1998. Inflammatory large bowel disease in immunodeficient rats naturally and experimentally infected with *Helicobacter bilis*. *Vet Pathol* 35(3): 202–208.

Halvorson, I. et al. 1992. Computer interface and software analysis for multi-channel, on-line temperature recording to any microcomputer. *Comput Biol Med* 22(1–2): 83–95.

Harkness, JE. et al., editors. 2010. *Harkness and Wagner's Biology and Medicine of Rabbits and Rodents*, 5th ed. Ames (IA): Blackwell Publishing, pp. 163–165.

Hawk, CT. et al. 2005. *Formulary for Laboratory Animals*, 3rd ed. Ames (IA): Blackwell Publishing.

Heavner, JE, Cooper, DE. 2008. Pharmacology of analgesics, In: Fish RE, Danneman PJ, Brown M, Karas AZ, editors. *Anesthesia and Analgesia in Laboratory Animals*, 2nd ed. London (UK): Academic Press, pp. 97–123.

Hickman, DL, Swan, M. 2010. Use of a body condition score technique to assess health status in a rat model of polycystic kidney disease. *J Am Assoc Lab Sci* 49(2): 155–159.

Hill, WA. et al. 2005. Use of permethrin eradicated the tropical rat mite (*Ornithonyssus bacoti*) from a colony of mutagenized and transgenic mice. *Contemp Top Lab Anim* 44(5): 31–34.

Hill, WA. et al. 2006. Efficacy and safety of topical selamectin to eradicate pinworm (*Syphacia* spp.): Infections in rats (*Rattus norvegicus*) and mice (*Mus musculus*). *J Am Assoc Lab Anim Sci* 45(3): 23–26.

Holder, DS. 1992. Effects of urethane, alphaxolone/alphadolone, or halothane with or without neuromuscular blockade on survival during repeated episodes of global ischemia in the rat. *Lab Anim* 26: 107–113.

Holman, JM, Jr, Saba, TM. 1988. Effect of bacterial sepsis on gluco-neogenic capacity in the Rat. *J Surg Res* 45(2): 167–175.

Holzer, P. et al. 2001. Estimation of acute flurbiprofen and ketoprofen toxicity in rat gastric mucosa at therapy-relevant doses. *Inflamm Res* 50(12): 602–608.

Hong, CC, Ediger RD. 1978. Chronic necrotizing mastitis in rats caused by *Pastereulla pneumotropica*. *Lab Anim Sci* 28(3): 317–320.

Hook, GE. 1991. Alveolar proteinosis and phospholipidoses of the lungs. *Toxicol Pathol* 19(4 Pt 1): 482–513.

Hu, C. et al. 1992. Fentanyl and medetomidine anaesthesia in the rat and its reversal using atipamazole and either nalbuphine or butorphanol. *Lab Anim* 26(1): 15–22.

Hughes, WT, Killmar, J. 1996. Monodrug efficacies of sulfonamides in prophylaxis for *Pneumocystis carinii* pneumonia. *Antimicrob Agents Chemother* 40(4): 962–965.

International Association for the Study of Pain. [Internet]. 1979. IASP pain terminology. [Cited 23 Oct 2011]. Available at: www.iasp-pain.org

Icenhour, CR. et al. 2001. Widespread occurrence of *Pneumocystis carinii* in commercial rat colonies detected using targeted PCR and oral swabs. *J Clin Microbiol* 39(10): 3437–3441.

Icenhour, CR. et al. 2006. Competitive coexistence of two *Pneumocystis* species. *Infect Genet Evol* 6(3): 177–186.

Ike, F. et al. 2007. Lymphocytic choriomeningitis infection unde-tected by dirty-bedding sentinel monitoring and revealed after embryo transfer of an inbred strain derived from wild mice. *Comp Med* 57(3): 272–281.

Isabel, J. et al. 2008. Anesthesia and other considerations for in vivo imaging of small animals. *ILAR J* 49(1): 17–26.

Ishih, A. et al. 1992. Differential establishment and survival of *Hymenolepis diminuta* in syngeneic and outbred rat strains. *J Helminthol* 66(2): 132–136.

Jacoby, RO. et al. 1991. Persistent rat parvovirus infection in indi-vidually housed rats. *Arch Virol* 117(3–4): 193–205.

Jacoby, RO. et al. 1996. Rodent parvovirus infections. *Lab Anim Sci* 46(4): 370–380.

Jang, HS. et al. 2009. Effects of propofol administration rates on cardiopulmonary function and anaesthetic depth during anaes-thetic induction in rats. *Vet Anaesth Analg* 36(3): 239–245.

Jang, HS. et al. 2009. Evaluation of the anaesthetic effects of medetomidine and ketamine in rats and their reversal with atipamezole. *Vet Anaesth Analg* 36(4): 319–327.

Jang, HS, Lee, MG. 2009. Atipamezole changes the antinociceptive effects of butorphanol after medetomidine-ketamine anaesthesia in rats. *Vet Anaesth Analg* 36(6): 591–596.

Jenkins, WL. 1987. Pharmacologic aspects of analgesic drugs in animals: An overview. *J Am Vet Med Assoc* 191(10): 1231–1240.

Johansen, O. et al. 1994. Increased plasma glucose levels after hypnorm anaesthesia, but not after pentobarbital anaesthesia in rats. *Lab Anim* 28(3): 244–248.

Johnson-Delaney, CA, Harrison, LR. 1996. Small rodents. In: Johnson-Delaney CA, Harrison LR, editors. *Exotic Companion Medicine Handbook for Veterinarians.* Lake Worth (FL): Wingers Publishing.

Jones, TC. et al. 1985. *Digestive System: Monographs on Pathology of Laboratory Animals.* New York: Springer-Verlag.

Jones, TC. et al. 1985. *Respiratory System: Monographs on Pathology of Laboratory Animals.* New York: Springer-Verlag.

Jury, J. et al. 2005. Eradication of *Helicobacter* spp. from a rat breeding colony. *Contemp Top Lab Anim Sci* 44(4): 8–11.

Kajiwara, N. et al. 1996. Vertical transmission to embryo and fetus in maternal infection with rat virus (RV). *Exp Anim* 45(3): 239–244.

Karwacki, Z. et al. 2001. General anaesthesia in rats undergoing experiments on the central nervous system. *Folia Morphol (Warsz)* 60(4): 235–242.

Kashimoto, A. et al. 1997. The minimum alveolar concentration of sevoflurane in rats. *Eur J Anaesthesiol* 14: 359–361.

Kelaher, J. et al. 2005. An Outbreak of rat mite dermatitis in an animal research facility. *Cutis* 75(5): 282–286.

Kennedy, HS. et al. 2011. A method for removing the brain and spinal cord as one unit from adult mice and rats. *Lab Anim (NY)* 40(2): 1.

Kerton, A, Warden, P. 2006. Review of successful treatment for helicobacter species in laboratory mice. *Lab Anim* 40(2): 115–122.

Kharasch, ED. et al. 1995. Clinical sevoflurane metabolism and disposition: 1. Sevoflurane and metabolite pharmacokinetics. *Anesthesiology* 82: 1369–1378.

King, CS, Miller, RT. 1997. Fatal perforating intestinal ulceration attributable to flunixin meglumine overdose in rats. *Lab Anim Sci* 47(2): 205–208.

Kitagaki, M. et al. 2003. Auricular chondritis in young ear-tagged Crj: CD(SD)IGS rats. *Lab Anim* 37(3): 249–253.

Kizelshteyn, G. et al. 1992. Enhancement of bupivacaine sensory blockade of rat sciatic nerve by combination with phenol. *Anesth Analg* 74(4): 499–502.

Kohn, DF, Clifford CB. 2002. Biology and diseases of rats. In: Fox JG, Anderson LC, Loew FM, Quimby FW, editors. *Laboratory Animal Medicine*, 2nd ed. San Diego (CA): Academic Press, pp. 121–165.

Kondo, S. et al. 1998. Elimination of an infestation of rat fur mites (*Radfordia ensifera*) from a colony of Long-Evans rats, using the micro-dot technique for topical administration of 1% ivermectin. *Contemp Top Lab Anim Sci* 37(1): 58–61.

Koppel, J. et al. 1982. Changes in blood sugar after administration of xylazine and adrenergic blockers in the rat. *Vet Med (Praha)* 27(2): 113–118.

Korezeniewska-Ryicka, I, Plaznik, A. 1998. Analgesic effect of antidepressant drugs. *Pharmacol Biochem Behav* 59(2): 331–338.

Kostomitsopoulos, N. et al. 2007. Eradication of *Helicobacter bilis* and *H. hepaticus* from infected mice by using a medicated diet. *Lab Anim (NY)* 36(5): 37–40.

Koszdin, KL, DiGiacomo, RF. 2002. Outbreak: Detection and investigation. *Contemp Top Lab Anim Sci* 41(3): 18–27.

Krinke, G. 1983. Spinal radiculoneuropathy in aging rats: Demyelination secondary to neuronal dwindling? *Acta Neuropathol* 59(1): 63–69.

Kubisch, HM, Gómez-Sánchez, EP. 1999. Embryo transfer in the rat as a tool to determine genetic components of the gestational environment. *Lab Anim Sci* 49(1): 90–94.

Kuiper, B. et al. 1997. Ophthalmologic examination in systemic toxicity studies: An overview. *Lab Anim* 31(2): 177–183.

Kurisu, K. et al. 1996. Sequential changes in the Harderian gland of rats exposed to high intensity light. *Lab Anim Sci* 46(1): 71–76.

Kurth, A. et al. 2008. Rat-to-elephant-to-human transmission of cowpox virus. *Emerg Infect Dis* 14(4): 670–671.

La Regina, M. et al. 1992. Transmission of sialodacryoadenitis virus (SDAV) from infected rats to rats and mice through handling, close contact, and soiled bedding. *Lab Animal Sci* 42(4): 344–346.

Larsen, HH. et al. 2002. Development of a rapid real-time PCR assay for quantitation of *Pneumocystis carinii* f. sp. carinii. *J Clin Microbiol* 40(8): 2989–2993.

Larsson, JE, Wahlström, G. 1998. The influence of age and administration rate on the brain sensitivity to propofol in rats. *Acta Anaesthesiol Scand* 42(8): 987–994.

Lawson, GW. et al. 2005. Vitamin E as a treatment for ulcerative dermatitis in C57BL/6 mice and strains with a C57BL/6 background. *Contemp Top Lab Anim Sci* 44(3): 18–21.

Lee, IH. et al. 2008. Spatial memory is intact in aged rats after propofol anesthesia. *Anesth Analg* 107(4): 1211–1215.

Liang, SC. et al. 1995. Comparative severity of respiratory lesions of sialodacryoadenitis virus and sendai virus infections in LEW and F344 rats. *Vet Pathol* 32(6): 661–667.

Liles, JH, Flecknell, PA. 1992. The effects of buprenorphine, nalbuphine and butorphanol alone or following halothane anesthesia on food and water consumption and locomotor movement in rats. *Lab Anim* 26(3): 180–189.

Liles, JH, Flecknell, PA. 1992. The use of nonsteroidal inflammatory drugs for the relief of pain in laboratory rodents and rabbits. *Lab Anim* 26: 241–255.

Linden, RD. et al. 2000. A laryngoscope designed for intubation of the rat. *Contemp Top Lab Anim Sci* 39(2): 40–42.

Livingston, RS. et al. 2010. *Pneumocystis carinii* infection causes lung lesions indistinguishable from those previously attributed to rat respiratory virus. Abstracts presented at the 61st AALAS National Meeting, Atlanta, GA, 10–14 Oct 2010. *J Am Assoc Lab Anim Sci* 49(5): 683.

Luger, TJ. et al. 2005. The effect of ciprofloxacin and gentamicin on spinal morphine-induced antinociception in rats. *Basic Clin Pharmacol Toxicol* 96: 366–374.

Maekawa, A, Odashima, S. 1975. Spontaneous tumors in ACI/N rats. *J Natl Cancer Inst* 55: 1437–1445.

Macy, JD, Jr. et al. 1996. Reproductive abnormalities associated with a coronavirus infection in rats. *Lab Anim Sci* 46(1): 129–132.

Mahabir, E. et al. 2008. Rodent and germplasm trafficking: Risks of microbial contamination in a high-tech biomedical world. *ILAR J* 49(3): 347–355.

Mahabir, E. et al. 2004. Mouse antibody production test: Can we do without it? *J Virol Methods* 120(2): 239–245.

Majeed, SK, Zubaidy, AJ. 1982. Histopathological lesions associated with *Encephalitozoon cuniculi* (nosematosis) infection in a colony of Wistar rats. *Lab Anim* 16(3): 244–247.

Makino, S. et al. 1972. An epizootic of sendai virus infection in a rat colony. *Exp. Anim* 22(4): 275–280.

Mangram, AJ. et al. 1999. Guideline for prevention of surgical site infection. *Infect Control Hosp Epidemiol* 20(4): 250–278.

Matson, DJ. et al. 2007. Inflammation-induced reduction of spontaneous activity by adjuvant: A novel model to study the effect of analgesics in rats. *J Pharmacol Exp Ther* 320(1): 194–201.

McCarthy, RJ. et al. 1998. Antinociceptive potentiation and attenuation of tolerance by intrathecal co-infusion of magnesium sulfate and morphine in rats. *Anesth Analg* 86: 830–836.

McCune, WJ. et al. 1982. Type II collagen-induced auricular chondritis. *Arthritis Rheum* 25(3): 266–273.

McEwen, BJ, Barsoum, NJ. 1990. Auricular chondritis in Wistar rats. *Lab Anim* 24(3): 280–283.

McKeon, GP. et al. 2011. Analgesic effects of tramadol, tramadol-gabapentin, and buprenorphine in an incisional model of pain in rats (*Rattus norvegicus*). *J Am Assoc Lab Anim Sci* 50(2): 192–197.

McKisic, MD. et al. 1995. A nonlethal rat parvovirus infection suppresses rat T lymphocyte effector functions. *J Immunol* 155(8): 3979–3986.

Meingassner, JG. 1991. Sympathetic auricular chondritis in rats: A model of autoimmune disease? *Lab Anim* 25(1): 68–78.

Mitsumori, K. et al. 1981. An ultrastructural study of spinal nerve roots and dorsal root ganglia in aging rats with spontaneous radiculoneuropathy. *Vet Pathol* 18(6): 714–726.

Mitsumori, K, Elwell, MR. 1988. Proliferative lesions in the male reproductive system of F344 rats and B6C3F1 mice: Incidence and classification. *Environ Health Perspect* 77: 11–21.

Morris, TH. 1995. Antibiotic therapeutics in laboratory animals. *Lab Anim* 29(1): 16–36.

Morrow, DT. et al. 1977. Poditis in the rat as a complication of experiments in exercise physiology. *Lab Anim Sci* 27(5 Pt 1): 679–681.

Muir, W. et al. 2007. *Handbook of Veterinary Anesthesia*, 4th ed. Philadelphia: Mosby Elsevier.

National Research Council. 2010. *Guide for the Care and Use of Laboratory Animals.* Washington, D.C.: National Academies Press.

National Research Council. 1991. *Infectious Diseases of Mice and Rats. A Report of the Institute of Laboratory Animal Resources Committee on Infectious Diseases of Mice and Rats.* Washington, D.C.: National Academies Press.

Nolen, RS. 2001. The responsibility to laboratory animals. *J Am Vet Med Assoc* 219(12): 1659.

Oge, H. et al. 2000. The effect of doramectin, moxidectin and neto-bimin against natural infections of *Syphacia muris* in rats. *Vet Parasitol* 88(3–4): 299–303.

Ohmura, S. et al. 2001. Systemic toxicity and resuscitation in bupivacaine-, levobupivacaine-, or ropivacaine-infused rats. *Anesth Analg* 93(3): 743–748.

Ordodi, VL. et al. 2006. Improved electrodes for electrical defibrilla-tion of rats. *J Am Assoc Lab Anim Sci* 45(6): 54–57.

Orliaguet, G. et al. 2001. Minimum alveolar concentration of volatile anesthetics in rats during posnatal maturation. *Anesthesiology* 95: 734–739.

Otto, G, Franklin, CL. 2006. Medical management and diagnos-tic approaches. In: Suckow MA, Weisbroth SH, Frankklin CL, editors. *The Laboratory Rat,* 2nd ed. Burlington (MA): Elsevier, pp. 547–564.

Paprottka, PM. et al. 2010. Non-invasive contrast enhanced ultra-sound for quantitative assessment of tumor microcirculation: Contrast mixed mode examination vs. only contrast enhanced ultrasound examination. *Clin Hemorheol Microcirc* 46(2): 149–158.

Parenteau, H. et al. 2009. Isoflurane does not alter the brain distri-bution of reference compounds and new chemical entities in the rat. Abstracts presented at the 60th AALAS National Meeting, Denver, CO, 8–12 Nov 2009. *J Am Assoc Lab Anim Sci* 48(5): 619.

Park, BY. et al. 2010. Antinociceptive effect of memantine and mor-phine on vincristine-induced peripheral neuropathy in rats. *Korean J Pain* 23(3): 179–185.

Paster, EV. et al. 2009. Endpoints for mouse abdominal tumor mod-els: Refinement of current criteria. *Comp Med* 59: 234–241.

Peace, TA. et al. 2001. Effects of caging type and animal source on the development of foot lesions in Sprague Dawley rats (*Rattus norvegicus*). *Contemp Top Lab Anim Sci* 40(5): 17–21.

Pearce, BD. et al. 1996. Persistent dentate granule cell hyperexcitability after neonatal infection with lymphocytic choriomeningitis virus. *J Neurosci* 6(1): 220–228.

Penderis, J, Franklin, RJ. 2005. Effects of pre-versus post-anaesthetic buprenorphine on propofol-anaesthetized rats. *Vet Anaesth Analg* 32(5): 256–260.

Percy, DH, Barthold, SW. 2007. Rat. In: *Pathology of Laboratory Rodents and Rabbits*, 3rd ed. Ames (IA): Blackwell, pp. 125–177.

Peters, BS. et al. 2010. The renin-angiotensin system as a primary cause of polyarteritis nodosa in rats. *J Cell Mol Med* 14(6A): 1318–1327.

Peterson, NC. 2008. From bench to cageside: Risk assessment for rodent pathogen contamination of cells and biologics. *ILAR J* 49(3): 310–315.

Pickering, RG, Pickering, CE. 1984. The effect of repeated reproduction on the Incidence of pituitary tumours in Wistar Rats. *Lab Anim* 18(4): 371–378.

Pickering, RG, Pickering, CE. 1984. The effect of diet on the incidence of pituitary tumours in female Wistar rats. *Lab Anim* 18(3): 298–314.

Pinkert, CA. 2002. *Transgenic Animal Technology: A Laboratory Handbook*, 2nd ed. San Diego (CA): Academic Press.

Postma, BH. et al. 1991. Cowpox-virus-like infection associated with rat bite. *Lancet* 337(8743): 733–734.

Prado, WA, Machado Filho, EB. 2002. Antinociceptive potency of aminoglycoside antibiotics and magnesium chloride: A comparative study on models of phasic and incisional pain in rats. *Braz J Med Biol Res* 35: 395–403.

Pritchett-Corning, KR. et al. 2009. Contemporary prevalence of infectious agents in laboratory mice and rats. *Lab Anim* 43(2): 165–173.

Pritchett, KR, Johnston, NA. 2002. A review of treatments for the eradication of pinworm infections from laboratory rodent colonies. *Contemp Top Lab Anim Sci* 41(2): 36–46.

Pullium, JK. et al. 2004. Rodent vendor apparent source of mouse parvovirus in sentinel mice. *Contemp Top Lab Anim* 43(4): 8–11.

Reeves, PG. et al. 1993. AIN-93 purified diets for laboratory rodents: Final report of the American Institute of Nutrition ad hoc writing committee on the reformulation of the AIN-76A rodent diet. *J Nutr* 123: 1939–1951.

Research Animal Diagnostic Laboratory, University of Missouri. [Internet]. 2011. MFI2: Next generation serology testing. [Cited 23 Oct 2011]. Available at: www.radil.missouri.edu

Ricart Arbona RJ. et al. 2010. Treatment and eradication of murine fur mites: III. Treatment of a large mouse colony with ivermectin-compounded feed. *J Am Assoc Lab Anim Sci* 49(5): 633–637.

Rodrigues, DM. et al. 2005. Theiler's murine encephalomyelitis virus in nonbarrier rat colonies. *Comp Med* 55(5): 459–464.

Roehl, AB. et al. 2011. Accidental renal injury by an external heating device during surgery in rats. *Lab Anim* 45: 45–49.

Rogers, MM. et al. 2003. Practical method for blood collection to prevent pathogen cross-contamination in suspected MPV-positive mouse colonies. Abstracts presented at the 54th AALAS National Meeting, Seattle, WA, 12–16 Nov 2003. *Contemp Top Lab Anim Sci* 42(4): 95.

Roughan, JV, Flecknell, PA. 2003. Evaluation of a short duration behaviour-based post-operative pain scoring system in rats. *Eur J Pain* 7(5): 397–406.

Rouleau, AMJ. et al. 1993. Decontamination of rat embryos and transfer to specific pathogen-free recipients for the production of a breeding colony. *Lab Anim Sci* 43(6): 611–615.

Rutala, WA. 1990. APIC guideline for selection and use of disinfectants. *Am J Infect Contr* 18: 99–117.

Saito, N, Kawamura, H. 1999. The incidence and development of periarteritis nodosa in testicular arterioles and mesenteric arteries of spontaneously hypertensive rats. *Hypertens Res* 22(2): 105–112.

Samsamshariat, SA, Movahed, MR. 2005. Using a 0.035-in. straight-tip wire and a small infant laryngoscope for safe and easy endotracheal intubations in rats for cardiovascular research. *Cardiovasc Revasc Med* 6(4): 160–162.

Schoeb, TR. et al. 1997. Pathogenicity of cilia-associated respiratory (CAR) bacillus isolates for F344, LEW, and SD rats. *Vet Pathol* 34(4): 263–270.

Schoeb, TR. et al. 2009. Mycoplasma pulmonis and lymphoma in bioassays in rats. *Vet Pathol* 46(5): 952–959.

Seely, JC, Hard, GC. 2008. Chronic progressive nephropathy (CPN) in the rat: Review of pathology and relationship to renal tumorigenesis. *J Tox Path* 21(4): 199–205.

Sharp, PE. et al. 2005. Developing and implementing a real-time webs-based information portal for biomedical research facilities. Abstracts presented at the 56th AALAS National Meeting, St. Louis, MO, 6–10 Nov 2005. *Contemp Top Lab Anim Sci* 44(4): 89.

Sharp, PE, Reed, MD. 2002. Palm OS applications facilitating biomedical research. Abstracts presented at the 53rd AALAS National Meeting, San Antonio, TX, 27–31 Oct 2002. *Contemp Top Lab Anim Sci* 41(4): 86.

Sharp, PE, Reed, MD. 2001. Use of a Palm OS® as a portable database and clinical management tool. Abstracts presented at the 52nd AALAS National Meeting, Baltimore, MD, 21–25 Oct 2001. *Contemp Top Lab Anim Sci* 40(4): 94.

Shearer, D. et al. 1973. Strain differences in the response of rats to repeated injections of pentobarbital sodium. *Lab Anim Sci* 23(5): 662–664.

Shek, WR. 2008. Role of housing modalities on management and surveillance strategies for adventitious agents of rodents. *ILAR J* 49(3): 316–325.

Shek, WR, Gaertner, DJ. 2002. Microbiological quality control for laboratory rodents and lagomorphs. In: Fox JG, Anderson LC, Loew FM, Quimby FW, editors. *Laboratory Animal Medicine*, 2nd ed. San Diego (CA): Academic Press, pp. 364–393.

Shibuya K. et al. 1986. Spontaneous occurrence of pulmonary foam cells in Fischer 344 Rats. *Jpn J Vet Sci* 48(2): 413–417.

Sinclair, MD. 2003. A review of the physiological effects of 2-agonists related to the clinical use of medetomidine in small animal practice. *Can Vet J* 44(11): 885–897.

Snijdelaar, DG. et al. 2005. Effects of pre-treatment with amantadine on morphine induced antinociception during second phase formalin responses in rats. *Pain* 119(1–3): 159–167.

Sorden, SD, Castleman, WL. 1995. Virus-induced increases in bronchiolar mast cells in Brown Norway rats are associated with both local mast cell proliferation and increases in blood mast cell precursors. *Lab Invest* 73(2): 197–204.

Soulez, B. et al. 1991. Introduction of *Pneumocystis carinii* in a colony of SCID mice. *J Protozool* 38: 123S–125S.

Spikes, SE. et al. 1996. Comparison of five anesthetic agents administered intraperitoneally in the laboratory rat. *Contemp Top Lab Anim Sci* 35(2): 53–56.

Sprague, JE. et al. 1993. Pharmacological studies of centrifugation-induced analgesia. *Pharmacol Biochem Behav* 46: 911–915.

Stark, RA. et al. 1981. Blind oral tracheal intubation of rats. *J Appl Physiol* 51(5): 1355–1356.

Stark, DM. 2001. Wire-bottom versus solid-bottom rodent caging issues important to scientists and laboratory animal science specialists. *J Am Assoc Lab Anim Sci* 40(6): 11–14.

Steffrey, EP, Khursheed RM, 2007. Inhalation anesthetics. In: Tranquilli JW. et al. editors. *Lumb and Jones' Veterinary Anesthesia and Analgesia*, 4th ed. Ames (IA): Wiley-Blackwell, pp. 355–393.

Stokes, EL. et al. 2009. Reported Analgesic and anaesthetic administration to rodents undergoing experimental surgical procedures. *Lab Anim* 43(2): 149–154.

Stratmann, G. et al. 2009. Increasing the duration of isoflurane anesthesia decreases the minimum alveolar anesthetic concentration in 7-day-old but not in 60-day-old rats. *Anesth Analg* 109(3): 801–806.

Stringer, SK, Seligmann, BE. 1996. Effects of two injectable anesthetic agents on coagulation assays in the rat. *Lab Anim Sci* 45(4): 430–433.

Suaudeau, C. et al. 1993. Analgesic effects of antibiotics in rats. *Pharmacol Biochem Behav* 46(2): 361–364.

Suaudeau, C. et al. 1989. Antibiotics and morphinomimemetic injections prevent automutilation behavior in rats after dorsal rhizotomy. *Clin J Pain* 5: 177–181.

Suzuki, H. et al. 1996. Rederivation of mice by means of in vitro fertilization and embryo transfer. *Exp Anim* 45(1): 33–38.

Takenami, T. et al. 2004. Intrathecal mepivacaine and prilocaine are less neurotoxic than lidocaine in a rat intrathecal model. *Reg Anesth Pain Med* 29(5): 446–453.

Takenami, T. et al. 2009. Spinal procaine is less neurotoxic than mepivacaine, prilocaine and bupivacaine in rats. *Reg Anesth Pain Med* 34(3): 189–195.

Taylor, DK. et al. 2006. Lanolin as a treatment option for ringtail in transgenic rats. *J Am Assoc Lab Anim Sci* 45(1): 83–87.

Thigpen, JE. et al. 1989. The use of dirty bedding for detection of murine pathogens in sentinel mice. *Lab Anim Sci* 39(4): 324–327.

Thomas, J. et al. 2007. A review of large granular lymphocytic leukemia in Fischer 344 rats as an initial step toward evaluating the implication of the endpoint to human cancer risk assessment. *Toxicol Sci* 99(1): 3–19.

Thomasy, SM. et al. 2007. Comparison of opioid receptor binding in horse, guinea pig, and rat cerebral cortex and cerebellum. *Vet Anaesth Analg* 34(5): 351–358.

Totton, M. 1958. Ringtail in new-born Norway rats, a study of the effect of environmental temperature and humidity on incidence. *J Hyg (Lon)* 56(2): 190–196.

Tran, DQ, Lawson, D. 1986. Endotracheal intubation and manual ventilation of the rat. *Lab Anim Sci* 36(5): 540–541.

Tsujio, M. et al. 2009. Bone mineral analysis through dual energy X-ray absorptiometry in laboratory animals. *J Vet Med Sci* 71(11): 1493–1497.

Tully, TN Jr. 2009. Mice and rats. In: Mitchell MA, Tully TN Jr., editors. *Manual of Exotic Pet Practice*. St. Louis (MO): Saunders Elsevier, pp. 326–344.

Turner, CH. et al. 2001. Variability in skeletal mass, structure, and biomechanical properties among inbred strains of rats. *J Bone Miner Res* 16(8): 1532–1539.

Van Keuren, ML, Saunders, TL. 2004. Rederivation of transgenic and gene-targeted mice by embryo transfer. *Transgenic Res* 13(4): 363–371.

Van Vleet, JF, Ferrans, VJ. 1986. Myocardial diseases of animals. *Am J Pathol* 124(1): 98–178.

Veiga, AP. et al. 2004. Prevention by celecoxib of secondary hyperalgesia induced by formalin in rats. *Life Sci* 75(23): 2807–2817.

Vogler, GA. 2006. Anesthesia and analgesia. In: Suckow MA, Weisbroth SH, Franklin CL, editors. *The Laboratory Rat*, 2nd ed. Burlington (MA): Elsevier, Inc., pp. 627–664.

Vonderfecht, SL. et al. 1984. Infectious diarrhea of infant rats produced by a rotavirus-like agent. *J Virol* 52(1): 94–98.

Vonderfecht, SL, Schemmer, JK. 1993. Purification of the IDIR strain of group B rotavirus and identification of viral structural proteins. *Virol* 194(1): 277–283.

Watson, J. 2008. New building, old parasite: Mesostigmatid mites—an ever-present threat to barrier rodent facilities. *ILAR J* 49(3): 303–309.

Watson, J. et al. 2005. Successful rederivation of contaminated immunocompetent mice using neonatal transfer with iodine immersion. *Comp Med* 55(5): 465–469.

Waynforth, HB, Flecknell, PA. 1992. *Experimental and Surgical Technique in the Rat*. London (UK): Academic Press.

Wegener, A. et al. 2002. Frequency and nature of spontaneous age-related eye lesions observed in a 2-year inhalation toxicology study in rats. *Ophthalmic Res* 34(5): 281–287.

Weisbroth, SH. et al. 1999. Latent *Pneumocystis carinii* infection in commercial rat colonies: Comparison of inductive immunosuppressants plus histopathology, PCR, and serology as detection methods. *J Clin Microbiol* 37(5): 1441–1446.

Weksler, B. et al. 1994. A simplified method for endotracheal intubation in the rat. *J Appl Physiol* 76(4): 1823–1825.

Whary, MT, Fox, JG. 2004. Natural and experimental *Helicobacter* infections. *Comp Med* 54(2): 128–158.

Whary, MT, Fox, JG. 2006. Detection, eradication, and research implications of *Helicobacter* infections in laboratory rodents. *Lab Anim* 35(7): 25–36.

Whelan, G, Flecknell, PA. 1992. The assessment of depth of anaesthesia in animals and man. *Lab Anim* 26(3): 153–162.

Whelan, G, Flecknell, PA. 1994. The use of etorphine/methotrimeprazine and midazolam as an anaesthetic technique in laboratory rats and mice. *Lab Anim* 28(1): 70–77.

Wiersma, J, Kastelijn, J. 1996. Haematological, immunological and endocrinological aspects of chronic high frequency blood sampling in rats with replacement by fresh or preserved donor blood. *Lab Anim* 20(1): 57–66.

Willens, SL, Sproul, EE. 1938. Spontaneous cardiovascular disease in the rat: I. Lesions of the heart. *Am J Pathol* 14(2): 177–200.

Williams, DL. 2002. Ocular disease in rats: A review. *Vet Ophthalmol* 5(3): 183–191.

Wixson, SK. et al. 1987. A comparison of pentobarbital, fentanyl-droperidol, ketamine-xylazine and ketamine-diazepam anesthesia in adult male rats. *Lab Anim Sci* 37(6): 726–730.

Wixson, SK. et al. 1987. The effects of pentobarbital, fentanyl-droperidol, ketamine-xylazine and ketamine-diazepam on arterial blood ph, blood gases, mean arterial blood pressure and heart rate in adult male rats. *Lab Anim Sci* 37(6): 736–742.

Wixson, SK. et al. 1987. The effects of pentobarbital, fentanyl-droperidol, ketamine-xylazine and ketamine-diazepam on core and surface body temperature regulation in adult male rats. *Lab Anim Sci* 37(6): 743–749.

Wixson, SK. et al. 1987. The effects of pentobarbital, fentanyl-droperidol, ketamine-xylazine and ketamine-diazepam on noxious stimulus perception in adult male rats. *Lab Anim Sci* 37(6): 731–735.

Wolf, RF. et al. 1992. Magnetic resonance imaging using a clinical whole body system: An introduction to a useful technique in small animal experiments. *Lab Anim* 26(3): 222–227.

Wolfs, TF. et al. 2002. Rat-to-human transmission of cowpox infection. *Emerg Infect Dis* 8(12): 1495–1496.

Yabuuchi, K. et al. 2010. A diagnostic method for *Pneumocystis carinii* a causative agent of pneumonia in immunodeficient rats. *Exp Anim* 59(2): 261–267.

Yildiran, G. et al. 1997. Antinociception induced by verapamil and chloramphenicol in mice. *Biol Neonate* 72: 28–31.

Zegre Cannon, C. et al. 2010. Comparison of analgesic effects of tramadol, carprofen, or multimodal analgesia in rats undergoing ventral laparotomy. *J Am Assoc Lab Anim Sci* 49(5): 692–693.

chapter 5

Adam, RJ. 2002. Techniques of experimentation. In: Fox JG, Anderson LC, Loew FM, Quimby FW, editors. *Laboratory Animal Medicine*, 2nd ed. San Diego (CA): Academic Press, pp. 1005–1045.

Anderson, PG. et al. 2006. Cardiovascular research. In: Suckow MA, Weisbroth SH, Franklin CL, editors. *The Laboratory Rat*, 2nd ed. Burlington (MA): Elsevier Academic Press, pp. 773–802.

Adriani, W. et al. 2003. The spontaneously hypertensive-rat as an animal model of ADHD: Evidence for impulsive and non-impulsive subpopulations. *Neurosci Biobehav Rev* 27(7): 639–651.

Artwohl, J. et al. 2009. Extreme susceptibility of African naked mole rats (*Heterocephalus glaber*) to experimental infection with herpes simplex virus type 1. *Comp Med* 59(1): 83–90.

Artwohl, J. et al. 2002. Naked mole-rats: Unique opportunities and husbandry challenges. *Lab Anim (NY)* 31(5): 32–36.

Arzadon J. [Internet]. 2011. Adaptation of submanidbular blood collection in rats. [Cited 23 Oct 2011]. Available at: www.medipoint.com

Bader, M. 2010. Rat models of cardiovascular diseases. *Methods Mol Biol* 597: 403–414.

Blasiak, T. et al. 2010. A new approach to detection of the bregma point on the rat skull. *J Neurosci Methods* 185(2): 199–203.

Bolliger, AP. 2004. Cytologic evaluation of bone marrow in rats: Indications, methods, and normal morphology. *Vet Clin Pathol* 33(2): 58–67.

Bonnichsen, M. et al. 2005. The welfare impact of gavaging laboratory rats. *Animal Welfare* 14(3): 223–227.

Boyd, M. et al. 2004. A stepwise surgical procedure to investigate the lymphatic transport of lipid-based oral drug formulations: cannulation of the mesenteric and thoracic ducts within the rat. *J Pharmacol Toxicol Methods* 49(2): 115–120.

Braintree Scientific, Inc. [Internet]. 2011. MD-PVC rat model and Koken Rat (experimental techniques training model). [Cited 3 Oct 2011]. Available at: www.braintreesci.com

Bryda, EC, Riley, LK. 2008. Multiplex microsatellite marker panels for genetic monitoring of common rat strains. *J Am Assoc Lab Anim Sci* 47(3): 37–41.

Buffenstein, R. 2005. The naked mole-rat: A new long-living model for human aging research. *J Gerontol A Biol Sci Med Sci* 60(11): 1369–1377.

Buffenstein, R. 2008. Negligible senescence in the longest living rodent, the naked mole-rat: Insights from a successfully aging species. *J Comp Physiol B* 178: 439–445.

Buffenstein, R, Yahav, S. 1991. Is the naked mole-rat *Heterocephalus glaber* an endothermic yet poikilothermic mammal? *J Therm Biol* 16: 227–232.

Buhalog, A. et al. 2010. A method for serial selective arterial catheterization and digital subtraction angiography in Rodents. *Am J Neuroradiol* 31(8): 1508–1511.

Burch, S. et al. 2007. Multimodality imaging for vertebral metastases in a rat osteolytic model. *Clin Orthop Relat Res* 454: 230–236.

Candelario-Jalil, E. et al. 2007. Cyclooxygenase inhibition limits blood-brain barrier disruption following intracerebral injection of tumor necrosis factor-α in the rat. *J Pharmacol Exp Ther* 323(2): 488–498.

Clark, GT. et al. 2011. Utilization of dried blood spots within drug discovery: Modification of a standard DiLab® AccuSampler® to facilitate automatic blood spot sampling. *Lab Anim* 45(4): 124–126.

Clarke R. 1996. Animal models of breast cancer: Their diversity and role in biomedical research. *Breast Cancer Res Treat* 39(1): 1–6.

Clarke, KA, Parker, AJ. 1986. A quantitative study of normal locomotion in the rat. *Physiol Behav* 38(3): 345–351.

Clough, G. 1991. Suggested guidelines for the housing and husbandry of rodents for aging studies. *Neurobiol Aging* 12(6): 653–658.

Coates, ME. 1991. Nutritional considerations in the production of rodents for aging studies. *Neurobiol Aging* 12(6): 679–682.

Coria-Avila, GA. et al. 2007. Cecum location in rats and the implications for intraperitoneal injections. *Lab Anim (NY)* 36(6): 25–30.

Corwin, RL. 2006. Bingeing rats: A model of intermittent excessive behavior? *Appetite* 46(1): 11–15.

Crawley, JN. 2003. Behavioral phenotyping of rodents. *Comp Med* 53(2): 140–146.

Daneshvar, H. et al. 2011. Cardiovascular assessment in radiotelemetry-implanted pregnant rats. *J Invest Surg* 24(1): 35–43.

Deerberg, F. 1991. Age-Associated versus husbandry-related pathology of aging rats. *Neurobiol Aging* 12(6): 659–662.

Deveney, AM. et al. 1998. A pharmacological validation of radiotelemetry in conscious, freely moving rats. *J Pharmacol Toxicol Methods* 40(2): 71–79.

de Wit, M. et al. 2001. Implantable device for intravenous drug delivery in the Rat. *Lab Anim* 35(4): 321–324.

Diehl, KH. et al. 2001. A good practice guide to the administration of substances and removal of blood, including routes and volumes. *J Appl Toxicol* 21: 15–23.

Doggrell, SA, Brown, L. 1998. Rat models of hypertension, cardiac hypertrophy and Failure. *Cardiovasc Res* 39(1): 89–105.

Edrey, YH. et al. 2011. Successful Aging and sustained good health in the naked mole rat: A long-lived mammalian model for biogerontology and biomedical research. *ILAR J* 52(1): 41–53.

Faith, RE. et al. 1997. The cotton rat in biomedical research. *Lab Anim Sci* 47(4): 337–345.

Faure, L. et al. 2006. Evaluation of a surgical procedure to measure drug biliary excretion of rats in regulatory safety studies. *Fundam Clin Pharmacol* 20(6): 587–593.

Festing, MF. 1991. Genetic quality control of laboratory animals used in aging studies. *Neurobiol Aging* 12(6): 673–677.

Fine, J. et al. 1986. Annotated bibliography on uncommonly used laboratory animals: Mammals. *ILAR News* 29(4): 4–10.

Fitzgerald, AL. et al. 2003. Development of a quantitative method for evaluation of the electroencephalogram of rats by using radiotelemetry. *Contemp Top Lab Anim Sci* 42(1): 40–45.

Fraser, TB. et al. 2001. Comparison of telemetric and tail-cuff blood pressure monitoring in adrenocorticotrophic hormone-treated rats. *Clin Exp Pharmacol Physiol* 28(10): 831–835.

Gabir, AA. et al. 2007. Alterations in cytoskeletal and immune function-related proteome profiles in whole rat lung following intratracheal instillation of heparin. *Respir Res* 8: 36.

Garner, D. et al. 1988. A new method for direct measurement of systolic and diastolic pressures in conscious rats using vascular-access ports. *Lab Anim Sci* 38: 205–207.

Gianutsos, G. 1997. Accumulation of manganese in rat brain following intranasal administration. *Fundamen Appl Toxicol* 37(2): 102–105.

Gohd, R. et al. 1974. A simple prefabricated electrode connector unit for chronic rat EEG studies. *Physiol Behav* 12(6): 1097–1099.

Gottesmann, C. et al. 1977. Polygraphic recording of the rat using miniaturised telemetry equipment. *Physiol Behav* 18(2): 337–340.

Greinke, DH, Fortin, D. 2011. Technique for intracolonic administration in rats for pharmacokinetic studies. Abstracts presented at the 62nd AALAS National Meeting, San Diego, CA, 2–6 Oct 2010. *J Am Assoc Lab Anim Sci* 50(5): 748.

Guazzo, EP. et al. 1995. A rat spinal subarachnoid continuous infusion device. *J Clin Neurosci* 2(4): 339–344.

Hadd, S. et al. 2011. Needle-free injections via the subcutaneous route in rats, using the Bioject® device: An alternative to dosing preclinical species. Abstracts presented at the 62nd AALAS National Meeting, San Diego, CA, 2–6 Oct 2010. *J Am Assoc Lab Anim Sci* 50(5): 756.

Handler, JA. et al. 1994. Assessment of hepatobiliary function in vivo and ex vivo in the rat. *J Pharmacol Toxicol Methods* 31: 11–19.

Harkness, JE. et al. 2010. *Harkness and Wagner's Biology and Medicine of Rabbits and Rodents*, 5th ed. Ames (IA): Blackwell Publishing, pp. 163–165.

Harvard Apparatus. [Internet]. 2011. Rodent vascular access ports with pre-attached catheters. [Cited 18 Nov 2011]. Available at: www.harvardapparatus.com

Hauss, D. et al. 1998. Chronic collection of mesenteric lymph from conscious, tethered rats. *Contemp Top Lab Anim Sci* 37(3): 56–58.

Hawk, CT. et al. 2005. *Formulary for Laboratory Animals*, 3rd ed. Ames (IA): Blackwell Publishing.

Hazzard, DG. 1991. Relevance of the rodent model to human aging studies. *Neurobiol Aging* 12(6): 645–649.

Hem, A. et al. 1998. Saphenous vein puncture for blood sampling of the mouse, rat hamster, gerbil, guinea pig, ferret and mink. *Lab Anim* 32: 364–368.

Hettinger, PC. et al. 2011. Long-term vascular access ports as a means of sedative administration in a rodent fMRI survival model. *J Neurosci Methods* 200(2): 106–112.

Huetteman, DA, Bogie, H. 2009. Direct blood pressure monitoring in laboratory rodents via implantable radio telemetry. *Methods Mol Biol* 573: 57–73.

Hull, RM. 1995. Guideline limit volumes for dosing animals in the preclinical stage of safety evaluation. *Hum Exp Toxicol.* 14(3): 305–307.

Ionac, M. 2003. One technique, two approaches, and results: Thoracic duct cannulation in small laboratory animals. *Microsurgery* 23(3): 239–245.

Instech Laboratories. [Internet]. 2011. Automated Blood Sampler. [Cited 20 Nov 2011]. Available at: www.instechlabs.com

Instech Laboratories. [Internet]. 2011. Vascular Access Harness™ for Rats. [Cited 20 Nov 2011]. Available at: www.instechlabs.com

Jackson, RK. 1997. Unusual laboratory rodent species: Research uses, care, and associated biohazards. *ILAR J* 38(1): 13–21.

Jarvis, JU. 1981. Eusociality in a mammal: Cooperative breeding in naked mole-rat colonies. *Science* 212: 571–573.

Jarvis, JUM. 1991. Appendix: Methods for capturing, transporting, and maintaining naked mole-rats in captivity. In: Sherman PW, Jarvis JUM, Alexander RD, editors. *The Biology of the Naked Mole-Rat*. Princeton (NJ): Princeton University Press, pp. 467–483.

Jarvis, JUM. 1991. Reproduction of naked mole-rats. In: Sherman PW, Jarvis JUM, Alexander RD, editors. *The Biology of the Naked Mole-Rat*. Princeton (NJ): Princeton University Press, pp. 405–411.

Jekl, V. et al. 2005. Blood sampling from the cranial vena cava in the Norway rat (*Rattus norvegicus*). *Lab Anim* 39(2): 236–239.

Jobe, PC. et al. 1995. The genetically epilepsy-prone rat (GEPR). *Ital J Neurol Sci* 16(1–2): 91–99.

Johnson-Delaney, CA, Harrison, LR. 1996. Small rodents. In: Johnson-Delaney CA, Harrison LR, editors. *Exotic Companion Medicine Handbook for Veterinarians*. Lake Worth (FL): Wingers Publishing.

Johnson, K. 2008. Introduction to rodent cardiac imaging. *ILAR J* 49(1): 27–34.

Kamendi, HW. et al. 2010. Combining radio telemetry and automated blood sampling: A novel approach for pharmacology and toxicology studies. *J Pharmacol Toxicol Methods* 62(1): 30–39.

Kerler, R, Rabes, HM. 1994. Rat tumor cytogenetics: A critical evaluation of the literature. *Crit Rev in Oncog* 5: 271–295.

Kiianmaa, K. et al. 1991. Determinants of alcohol preference in the AA and ANA rat lines selected for differential ethanol intake. *Alcohol Alcohol (Suppl)* 1: 115–120.

Klimczak, A. et al. 2007. Donor-origin cell engraftment after intraosseous or intravenous bone marrow transplantation in a rat model. *Bone Marrow Transplant* 40: 373–380.

Knutson, DW. et al. 1977. Association and dissociation of aggregated IgG from rat peritoneal macrophages. *J Exp Med* 145: 1368–1381.

Koch MA. 2006. Experimental modeling and research methodology. In: Suckow MA, Weisbroth SH, Franklin CL, editors. *The Laboratory Rat*, 2nd ed. Burlington (MA): Elsevier Academic Press, pp. 587–625.

Kuwahara, M. et al. 1994. Power spectral analysis of heart rate variability as a new method for assessing autonomic activity in the Rat. *J Electrocardiol* 27(4): 333–337.

Lebedev, LV. et al. 2004. Spatial characteristics of cisterna magna in rats and novel technique for puncture with a stereotactic manipulator. *B Exp Biol Med* 137(6): 635–638.

Letelier, ME. et al. 1985. Tolerance and detoxicating enzymes in *Octodon degus* and Wistar rats: A comparative study. *Comp Biochem Physiol C* 80(1): 195–198.

Letelier, ME. et al. 1984. Enhanced metabolism of morphine in *Octodon degus* compared to Wistar rats. *Gen Pharmacol* 15(5): 403–406.

Lewis, DN. et al. 2005. 2-Butoxyethanol female-rat model of hemolysis and disseminated thrombosis: X-ray characterization of osteonecrosis and growth-plate suppression. *Toxicol Pathol* 33(2): 272–282.

Lizio, R. et al. 2001. Oral endotracheal intubation of rats for intratracheal instillation and aerosol drug delivery. *Lab Anim* 35: 257–260.

Lombry, C. et al. 2002. Confocal imaging of rat lungs following intratracheal delivery of dry powders or solutions of fluorescent probes. *J Control Release* 83: 331–341.

LoPachin, RM. et al. 1981. An improved method for chronic catheterization of the rat spinal subarachnoid space. *Physiol Behav* 27(3): 559–561.

Lucilla, B. et al. 2004. Receptor-mediated endocytosis of biofilm-forming *Enterococcus faecalis* by rat peritoneal macrophages. *Indian J Med Res* 119(Suppl): 131–135.

Lukas, G. et al. 1971. The route of absorption of intraperitoneally administered compounds. *J Pharmacol Exp Ther* 178(3): 562–566.

Luo, YS. et al. [Internet]. 2000. Comparison of catheter lock solutions in rats: Vascular Access Harness™ for rats. [Cited 2 Nov 2011]. Available at: www.criver.com

Manor, D, Sadeh, M. 1989. Muscle fiber necrosis induced by intramuscular injection of drugs. *Br J Exp Path* 70: 457–462.

Mao, C. et al. 2007. Got pure blood in fetal rats? *Pediatr Hematol Oncol* 24(6): 457–460.

Martín Barrasa, JL. et al. 2008. Electrocardiographic changes in rats undergoing thoracic surgery under combined parenteral anesthesia. *Lab Anim (NY)* 37(10): 469–474.

Mason-Bright, TK. et al. 2011. Reduced rat usage through a comparison of catheter placement for serial blood sampling in rats. Abstracts presented at the 62nd AALAS National Meeting, San Diego, CA, 2–6 Oct 2011. *J Am Assoc Lab Anim Sci* 50(5): 757.

McComb, MM. et al. 2011. Long-term maintenance of chronic catheterized rats used in standard self-administration study designs. Abstracts presented at the 62nd AALAS National Meeting, San Diego, CA, 2-6 Oct 2011. *J Am Assoc Lab Anim Sci* 50(5): 804.

Messonier, SP. 1998. Recognizing common signs of illness in pet rodents. *Vet Med* 93: 981–987.

Mestre, C. et al. 1994. A method to perform direct transcutaneous intrathecal injection in rats. *J Pharmacol Toxicol Methods* 32(4): 197–200.

Morton, DB. et al. 2001. Refining procedures for the administration of substances: Report of the BVAAWF/FRAME/RSPCA/UFAW Joint Working Group on Refinement. *Lab Anim* 35: 1–41.

Munakata A. et al. 1995. Effects of dietary fiber on gastrointestinal transit time, fecal properties and fat absorption in Rats. *Tohoku J Exp Med* 176(4): 227–238.

Murphy, JC. et al. 1978. Hematologic and serum protein reference values of the *Octodon degus*. *Am J Vet Res* 39(4): 713–715.

Nadon, NL. 2006. Gerontology and age-associated lesions. In: Suckow MA, Weisbroth SH, Frankklin CL, editors. *The Laboratory Rat*, 2nd ed. Burlington (MA): Elsevier Academic Press, pp. 761–772.

National Centre for Replacement, Refinement and Reduction of Animals in Research. [Internet]. 2011. Rat. [Cited 23 Oct 2011]. Available at: www.nc3rs.org.uk

National Institutes of Health. 2002. *Methods and Welfare Considerations in Behavioral Research with Animals: Report of a National Institutes of Health Workshop.* Morrison AR, Evans HL, Ator NA, Nakamura RK, editor. NIH Publication No. 02-5083. Washington, D.C.: US Government Printing Office.

National Research Council. 2003. *The Guidelines for the Care and Use of Mammals in Neuroscience and Behavioral Research.* Washington, D.C.: National Academies Press.

National Toxicology Program. [Internet]. 2011. Toxicology/ Carcinogenicity. [Cited 23 Oct 2011]. Available at: http: //ntp. niehs.nih.gov/

NeuroDetective International. [Internet]. 2011. Protocol for rat sleep EEG. [Cited 25 Oct 2011]. Available at: www.ndineuroscience.com/

Nlasoro, EJ. 1991. Use of rodents as models for the study of 'normal aging': Conceptual and practical issues. *Neurobiol Aging* 12(6): 639–643.

O'Connor, TP. et al. 2002. Prolonged longevity in naked mole-rats: Age-related changes in metabolism, body composition and gastrointestinal function. *Comp Biochem Physiol A: Mol Integr Physiol* 133: 835–842.

Ohwada, K. et al. 1994. Reference values for blood chemistry in the cotton rat (*Sigmodon hispidus*). *Scand J Lab Anim Sci* 21(1): 29–31.

Oka, Y. et al. 2006. A reliable method for intratracheal instillation of materials to the entire lung in rats. *J Toxicol Pathol* 19: 107–109.

Olds, RJ, Olds, JR. 1988. *A Colour Atlas of the Rat: Dissection Guide.* London (UK): Wolfe Medical Publications.

Pagés, T. et al. 1993. A method for sampling representative muscular venous blood during exercise in rats. *Lab Anim* 27(2): 171–175.

Park, TJ. et al. 2008. Selective inflammatory pain insensitivity in the African naked mole-rat (*Heterocephalus glaber*). *PLoS Biol* 6(1): e13.

Parker, AJ, Clark, KA. 1990. Gait topography in rat locomotion. *Physiol Behav* 48: 41–47.

Patterson, D. et al. 2011. Automated pharmacokinetic sampling in the rat by both dried blood spot and liquid blood methods. Abstracts presented at the 62nd AALAS National Meeting, San Diego, CA, 2–6 Oct 2011. *J Am Assoc Lab Anim Sci* 50(5): 794–795.

Paulose, CS, Dakshinamurti, K. 1987. Chronic catheterization using vascular-access-port in rats: Blood sampling with minimal stress for plasma catecholamine determination. *Neurosci Methods* 22(2): 141–146.

Paxinos, G, Watson, C. 2007. *The Rat Brain in Stereotaxic Coordinates,* 6th ed. London: Academic Press.

Paxinos, G. et al. 1985. Bregma, lambda and the interaural midpoint in stereotaxic surgery with rats of different sex, strain and weight. *J Neurosci Methods* 13(2): 139–143.

Pereira, JE. et al. 2006. A comparison analysis of hindlimb kinematics during overground and treadmill locomotion in rats. *Behav Brain Res* 172: 212–218.

Pereira-Junior, PP. et al. 2010. Noninvasive method for electrocardiogram recording in conscious rats: Feasibility for heart rate variability analysis. *An Acad Bras Cienc* 82(2): 431–437.

Petty, C. 1982. *Research Techniques in the Rat.* Springfield (IL): Charles C Thomas Publisher.

Phelan JP. 1992. Genetic variability and rodent models of human aging. *Exp Gerontol* 27(2): 147–159.

Phelan, JP, Austed, SN. 1994. Selecting animal models of human beings: Inbred strains often exhibit less biological uniformity than F1 hybrids. *J Gerontol* 49(1): B1–B11.

Popp, MB, Brennan, MF. 1981. Long-term vascular access in the rat: Importance of asepsis. *Am J Physiol* 241(14): H606–H612.

Porter, D. et al. 1999. Acute inflammatory reaction in rats after intratracheal instillation of material collected from a nylon flocking plant. *J Toxicol Environ Health* 57: 25–45.

Prediger, RD. et al. 2011. The intranasal administration of 1-methyl-4-phenyl-1,2,3,6-tetrahydropyridine (MPTP): A new rodent model to test palliative and neuroprotective agents for Parkinson's disease. *Curr Pharm Des* 17(5): 489–507.

Provencher Bolliger, A. 2004. Cytologic evaluation of bone marrow in rats: Indications, methods, and normal morphology. *Vet Clin Pathol* 33: 58–67.

Rago, B. et al. 2011. Application of the dried spot sampling technique for rat cerebrospinal fluid sample collection and analysis. *J Pharm Biomed Anal* 55: 1201–1207.

Rescue Critters! [Internet]. 2011. Squeekums. [Cited 23 Oct 2011]. Available at: www.rescuecritters.com/

Robineau, P. 1988. A simple method for recording electrocardiograms in conscious, unrestrained rats. *J Pharmacol Methods* 19(2): 127–133.

Rodgers, CT. 1995. Practical aspects of milk collection in the rat. *Lab Anim* 29(4): 450–455.

Rolf, LL Jr. et al. 1991. Chronic bile duct cannulation in laboratory rats. *Lab Anim Sci* 41: 486–492.

Russo, J, Russo, I. 1996. Experimentally induced mammary tumors in rats. *Breast Cancer Res Treat* 39(1): 7–20.

Salauze, D, Cave, D. 1995. Choice of vehicle for three-month continuous intravenous toxicology studies in the rat: 0.9% saline versus 5% glucose. *Lab Anim* 29(4): 432–437.

Sambhi, MP, White, FN. 1960. The electrocardiogram of the normal and hypertensive rat. *Circ Res* 8: 129–134.

Sanvitto, GL. et al. 1987. A technique for collecting cerebrospinal fluid using an intraventricular cannula in rats. *Physiol Behav* 41(5): 523–524.

Schwartz, ES. et al. 2009. Persistent pain is dependent on spinal mitochondrial antioxidant levels. *J Neurosci* 29(1): 159–168.

Sebesteny, A. 1991. Necessity of a more standardized microbiological characterization of rodents for aging studies. *Neurobiol Aging* 12(6): 663–668.

Seluanov, A. et al. 2009. Hypersensitivity to contact inhibition provides a clue to cancer resistance of naked mole-rat. *Proc Natl Acad Sci* 106(46): 19352–19357.

Sharma, AK. et al. 2006. Development of a percutaneous cerebrospinal fluid collection technique in F-344 rats and evaluation of cell counts and total protein concentrations. *Toxicol Pathol* 34(4): 393–395.

Sikes, RS, Gannon, WI, Animal Care and Use Committee of the American Society of Mammalogists. 2011. Guidelines of the American Society of Mammalogists for the use of wild mammals in research. *J Mammal* 92(1): 235–253.

Simpson, J, Kelly, JP. 2011. The Impact of environmental enrichment in laboratory rats: Behavioural and neurochemical aspects. *Behav Brain Res* 222(1): 246–264.

Sørensen, DB. et al. 2001. The influence of strain on demand functions for water in rats (*Rattus norvegicus*). *Scand J Lab Anim Sci* 28(1): 1–9.

Steele, VE. et al. 1994. Preclinical efficacy evaluation of potential chemopreven tative agents in animal careinogenesis models: Methods and results from the NCI Chemoprevention Drug Development Program. *J Cell Biochem* Suppl 20: 32–54.

Stokes, AH. et al. 2011. Determination of drug concentrations using dried blood spots: Investigation of blood sampling and collection techniques in Crl: CD(SD) rats. *Lab Anim* 45(4): 109–113.

Størkson, RV. et al. 1996. Lumbar catheterization of the spinal subarachnoid space in the rat. *J Neurosci Methods* 65(2): 167–172.

Strake, JG. et al. 1996. Model of *Streptococcus pneumoniae* meningitis in adult rats. *Lab Anim Sci* 46(5): 524–529.

Šulla, I. et al. 2008. Harvesting of adult rat bone marrow derived stem cells expressing nestin. *Folia Veterinaria* 52(3–4): 174–180.

Swearengen, JR. 2005. *Biodefense: Research Methodology and Animal Models*. Boca Raton (FL): CRC Press.

Swindle, MM. et al. 2005. Vascular access port (VAP) usage in large animal species. *Contemp Top Lab Anim Sci* 44(3): 7–17.

Tanaka, E. et al. 2008. Real-time intraoperative assessment of the extrahepatic bile ducts in rats and pigs using invisible near-infrared fluorescent light. *Surgery* (1): 39–48.

Tang, X. et al. 2007. Sleep and EEG Spectra in rats recorded via telemetry during surgical recovery. *Sleep* 30(8): 1057–1061.

Tavassoli, M. et al. 1970. Studies on marrow histogenesis: 1. The site of choice for extramedullary marrow implants. *Proc Soc Exp Biol Med* 133(3): 878–881.

Tsai, JC. et al. 1992. An experimental animal model of adenovirus-induced ocular disease: The cotton rat. *Arch Ophthalmol* 110(8): 1167–1170.

Tully, TN, Mitchell, MA. 2001. *A Technician's Guide to Exotic Animal Care*. Lakewood (CO): American Animal Hospital Association Press.

Turner, PV. et al. 2011. Administration of substances to laboratory animals: Routes of administration and factors to consider. *J Am Assoc Lab Anim Sci* 50(5): 600–613.

Turner, PV. et al. 2011. Administration of substances to laboratory animals: Equipment considerations, vehicle selection, and solute preparation. *J Am Assoc Lab Anim Sci* 50(5): 614–627.

Vainio, H, Cardis, E. 1992. Estimating human cancer risk from the results of animal experiments: Relationship between mechanism dose-rate and dose. *Am J Ind Med* 21(1): 5–14.

van der Logt, JTM. 1991. Necessity of a more standardized virological characterization of rodents for aging studies. *Neurobiol Aging* 12(6): 669–672.

van Herck, H. et al. 2001. Blood sampling from the retro-orbital plexus, the saphenous vein and the tail vein in rats: Comparative effects on selected behavioural and blood variables. *Lab Anim* 35(2): 131–139.

van Herck, H. et al. 1992. Histological changes in the orbital region of rats after orbital bleeding. *Lab Anim* 26: 53–58.

van Wijk, H. et al. 2001. A novel bile duct cannulation method with tail cuff exteriorization allowing continuous intravenous infusion and enterohepatic recirculation. *Lab Anim* 35(4): 325–333.

Wang, B. et al. 2010. An improved method for sampling thoracic duct lymph fluid in rats. Abstracts presented at the 61st AALAS National Meeting, Atlanta, GA, 10–14 Oct 2011. *J Am Assoc Lab Anim Sci* 49(5): 744.

Wang, F. et al. 2004. Intranasal delivery of methotrexate to the brain in rats bypassing the blood-brain barrier. *Drug Delivery Technology* 4(1): 48–55.

Wang, YM, Reuning, RH. 1994. A comparison of two surgical techniques for preparation of rats with chronic bile duct cannulae for the investigation of enterohepatic circulation. *Lab Anim Sci* 44: 479–485.

Ward, LC, Battersby, KJ. 2009. Assessment of body composition of rats by bioimpedence spectroscopy: Validation against dual-energy X-ray absorptiometry. *Scand J Lab Anim Sci* 36(3): 253–261.

Waynforth, HB, Flecknell, PA. 1992. *Experimental and Surgical Technique in the Rat.* London (UK): Academic Press.

Weiss, J. et al. 2000. Collection of body fluids. In: Krinke GJ, editor. *The Laboratory Rat.* London (UK): Academic Press, pp. 485–510.

Wolf, LW. et al. 1996. A behavioral study of the development of hereditary cerebellar ataxia in the shaker rat mutant. *Behav Brain Res* 75: 67–81.

Wright, JW, Kem, MD. 1992. Sterotaxic atlas of the brain of *Octodon degus. J Morphol* 214(3): 299–320.

Yaksh, TL, Rudy, TA. 1976. Analgesia mediated by a direct spinal action of narcotics. *Science* 192 (4246): 1357–1358.

Xiao, J. 2007. A new coordinate system for rodent brain and variability in the brain weights and dimensions of different ages in the naked mole-rat. *J Neurosci Methods* 162(1–2): 162–170.

Xiao, J. et al. 2006. A stereotaxic atlas of the brain of the naked mole-rat (*Heterocephalus glaber*). *Neuroscience* 141(3): 1415–1435.

Xu, F. et al. 2009. An improved method for protecting and fixing the lumbar catheters placed in the spinal subarachnoid space of rats. *J Neurosci Methods* 183(2): 114–118.

index